Amateur Radio Technician

Tricks for Beginners to Master
Ham Radio Basics

*(Ace Your Amateur Radio Technician Class Test
With Ease)*

Annetta Russel

Published By **Simon Dough**

Annetta Russel

All Rights Reserved

Amateur Radio Technician: Tricks for Beginners to Master Ham Radio Basics (Ace Your Amateur Radio Technician Class Test With Ease)

ISBN 978-1-77485-910-0

No part of this guidebook shall be reproduced in any form without permission in writing from the publisher except in the case of brief quotations embodied in critical articles or reviews.

Legal & Disclaimer

The information contained in this ebook is not designed to replace or take the place of any form of medicine or professional medical advice. The information in this ebook has been provided for educational & entertainment purposes only.

The information contained in this book has been compiled from sources deemed reliable, and it is accurate to the best of the Author's knowledge; however, the Author cannot guarantee its accuracy and validity and cannot be held liable for any errors or omissions. Changes are periodically made to this book. You must consult your doctor or get professional medical advice before using any of the suggested remedies, techniques, or information in this book.

Upon using the information contained in this book, you agree to hold harmless the Author from and against any damages, costs, and expenses, including any legal fees potentially resulting from the application of any of the information provided by this guide. This disclaimer applies to any damages or injury caused by the use and application, whether directly or indirectly, of any advice or information presented, whether for breach of contract, tort, negligence, personal injury, criminal intent, or under any other cause of action.

You agree to accept all risks of using the information presented inside this book. You need to consult a professional medical practitioner in order to ensure you are both able and healthy enough to participate in this program.

TABLE OF CONTENTS

Chapter 1: Preparatory Steps 1

Chapter 2: Basic Regulations 6

Chapter 3: Radio Waves 13

Chapter 4: The Electronic Basics 19

Chapter 5: Antennas And Feed-Lines 25

Chapter 6: Modulation And Signals 31

Chapter 7: Tips For Safety 36

Chapter 8: Station Set-Up 42

Chapter 9: Operating Your Amateur Station 50

Chapter 10: Fcc Rules And Guidelines 64

Chapter 11: Electrical And Electronic Parts That Make Up The Amateur Radio Station 76

Chapter 12: Commission's Rules In Amateur Radio Station 122

Chapter 13: Amateur Radio Operation Procedures 159

Conclusion 184

Chapter 1: Preparatory Steps

For the purpose of preparing for this exam, a lot of dedication is required from the intended student. It is necessary to read every section in the study guide, then read each question, go through all answers, then read the explanation of the correct answer, then mark the correct answer.

What are Amateur Radio Operators? An individual who has been who is granted an amateur operator license and recognized by the FCC ULS database is recognized as an amateur radio operator. The acronym ULS is a reference to the Universal Licensing System. It is the main database that contains the license information for all radio broadcasting services whether they're TV or commercial.

What is Amateur Radio Station? Amateur radio, also called Ham Radio is a non-commercial radio, but it is a licensed radio service that uses assigned radio frequencies not associated with frequencies such as FM or AM. This allows radio enthusiasts to communicate in emergency situations. It includes all the necessary equipment to communicate radio. The equipment here can be a handheld radio.

It's a dream to have an own station on the radio, doesn't it? The benefits are huge, and so much that astronauts find it intriguing as they can use it to establish contact or even send photos. A lot of operators have DIYs setups that comprise Aduino's Raspberry Pi and other micro controllers. These could be used to build GPF trackers. These are installed in vehicles or other higher elevations to broadcast their location via Amateur radio. You could even fire up an Ham to be able to reach the distance of a drone if you discover that the Wi-Fi connection you have isn't strong enough.

What is the reason for Amateur Radio Service Exists? Amateur Radio Service Exist Amateur Radio is used for recreation purposes and also as an activity. It's not something that we owe the government, however, it is something we need to know, so we have to pay for an Amateur Radio License. While Amateur Radio can be viewed as a form of entertainment but it's not an enjoyable pastime however, the FCC does not view it as an activity for fun. As per the FCC and the US this is their perspective that is important and not any other's. According to the FCC, and in America as well, Amateur Radio Service is of no interest to the public, and it's neither a commercial, commercial or public radio service. The FCC does not refer to Amateur Radio Service anywhere in its rules as an activity that is purely for fun!

Someone who has an amateur license radio service is known as an "Amateur Operator" but not an "Amateur Radio Operator" because the license is not granted to operate Radio. One example of a radio operating licence is that of one that is the GMRS license. It is the Amateur Radio is primarily for experiments, research and development of radio communications art.

According to FCC part 47of the FCC's rules "a primary purpose, as defined in the rules that follow, continuing and expanding the capabilities of amateur radio improves the development of the art of radio". This ensures the enhancement of the amateur radio service with guidelines that provide for growth in communication abilities and technical aspects of radio technology. This means that the current pool of amateur radio service that is populated by trained operators is increased, and the capacity of technicians and experts in electronics are kept in check.

The FCC has three license classes available each license class offers you specific rights to use specific frequencies and modes. Be aware that the more advanced the class or license you have, the greater number of allowed frequency bands that you are allowed to utilize.

Another purpose for the radio amateur service is that in limited instances, it can be employed for emergency communications. The use of the service for any other purpose beyond the ones previously mentioned is not allowed.

A higher percentage of Hams are failing in this area, since they don't actively contribute to the advancement of science in the Radio as well as have inadequate technical education. Hams aren't supposed to be on air. If they wish to chat with their pals it should be done via their mobile phone, GMRS or CB. Activities like contesting, storm-sporting or POTA, as well as communication during public events, are prohibited" in the Amateur rules. They are viewed as an indecent violation of the rules.

Chapter 2: Basic Regulations

How Do I get My license?

* Search for "ARRL examination locations" to locate an exam location near you.

* Make the payment of $15. It is managed by ARRL as well as the FCC and the FCC as well.

• If the test is not passed the exam, you are able to take the test and pay a fee for it.

When you're successful then you can take your test to pass the General as well as the Amateur Extra for free.

If you pay close to everything included throughout this text, you'll have everything you need to pass the exam. The bolded items are on the test, take them in your mind.

Try your hand at the hamstudy.org for no cost before you take the test or buy the Hamstudy.org App for your smartphone.

Basic Electricity

Voltage is the term used in the electrical industry for electromotive force (EMF)

Volt is the basic unit of electromotive force. Volt is the unit of base of electromotive force.

A little over 12 Volts is about the voltage an electronic transceiver on the move will need

Current is the term used to describe the flow of electrons through the electric circuit. The electrical Current is expressed in amps (amps)

Direct Current is the name for the current that only flows in one direction. A excellent example of this is the car battery.

Alternating Current is the reverse of a current frequently. The power source that is in your home is an example of.

Frequency is a term that's used to define the frequency of second an alternating current turns in the opposite direction.

The term "resistance" is used to define the opposition to flow of current within the circuit. The fundamental measurement of resistance is called ohm.

Insulators are substances that have great resistance to heat, such as plastic or glass

Conductors are the materials which allow the flow of electricity well, such as copper and metal.

Power is the speed that electrical energy is utilized or generated. Electric power can be measured as Watts (W)

Ohm's Law

* E=I*R

* Energy (in the form of volts) is I (current amps) * Resistance (in Ohms)

If you want to be able to remember, that formula "VCR" or Volts = current*resistance

To Calculate Power (DC Circuit)

*P=E*I (P=I*E(PIE) That's my secret

*Power (in Watts) = Energy (in Volts)*I (Current in Amps)

* Or to think of that "PVC" also known as Power = Volts*Current.

Two formulas you should know since they will be tested on the test. A few examples of the types of questions that you will see in the test

* 13.8V * 10A=138Watts

* 90v/3A = 30 Ohms

The questions usually appear in the form of word problems. So you'll be able to comprehend it and then, using the formula solve the problem by changing the questions to the formulas provided above.

Electronics Metrics

They are the various measurements for the levels of Volts, Amps Ohms Hertz and so on.

* giga (G) (1billion), which means 1Gigahert equals 1,000 Megahertz
* Mega (M) (1million), that is, 1 Megahert equals 1,000 Kilohertz.
* Kilo (k) (1thousand), which means 1 Kilhertz equals 1,000 Hertz.
* Milli (m) (1thousandth), which means 1mV (millivolt) is one thousandth of the millivolt.
* Micro (u) (1millionth), employs the Greek symbol mu (u) to symbolize micro
* nano (n) (1billionth), which means 1nv (nanovolt) can be described as one millionth one voltage

*Pico (p) (1trillionth), which translates to 1Pohm (picoohm) also known as 1 trillionth Ohm

Decibel System

Decibels are often used to talk about ratios and power ratios, and power. They operate at an algorithmic scale and not linear scales. Here are some examples.

3.dB represents the appropriate amount of power boost from 5 10 watts to 5 watts or a 2:1 ratio (twice as strong)

The *- 6dB value is the appropriate amount of decrease in power from 12 watts down to 3 milliwatts, which is 4:1 ratio (4times less effective in this instance, because it's a negative value)

The 10dB value is the ideal amount of power to increase from 20 power to 200 watts or a ratio of 10:1 (10 times as strong)

Radio Waves

* Electromagnetic waves; transmits radio signals between the transmitting stations as well as the receiving stations, providing crystal clear signals

* Radio waves, which travel across space. These waves are composed of magnetic and electric fields.
* Hertz (Hz) is the unit used to measure frequency it is named for "Heinrich Hertz", it encompasses one cycle every second
* Abbreviation is RF, refers to Radio Frequencies of all types
* Radio Wave travels at the speed of light. That is around 300,000,000 meters per second. traveling through space.
* Wavelength is the word used for the distance that a radio wave travels in an entire period (one wave)
* Because radio waves move at the rate of light the wavelength decreases when frequencies rise and we then can use an equation to convert frequency into a wavelength.
*The wavelength of meters is 300 times the megahertz frequency.

It is the approximate wavelength that's frequently used to define the various frequency bands. For example, the 2 meters ham radio band operates between

144 and 148 MHz, i.e. 300 times 148 which equals 2.03meters.

Frequencies

The frequency spectrum includes almost all energy transference and encompasses sound waves.

* It is important to note the fact that one KiloHertz (kHz) equals 1000Hz, 1 MegaHertz(MHz) is 1000 kHz and 1 gigaHertz ((GHz) is 1000 MHz. The radio spectrum is divided into various sections.

* VLF; Very Low Frequency 30-Hz to 30kHz

* LF; Low Frequency from 30kHz to 300kHz

* MF: Medium Frequencies : 300MHz to 3MHz

*HF, High Frequency 3MHz to 30MHz

* VHF, Very High Frequency 30MHz to 300MHz

* UHF, Ultra High Frequency 300MHz to 3GHz

* SHF; Super High Frequency 3GHz to 30GHz

* EHF 30GHz and above

Chapter 3: Radio Waves

Radio Wave Characteristics

* Communication between VHF and UHF is typically line of sight, which means they are straight from the receiver to the transmitter and, therefore, are not able to traverse the Earth and therefore aren't reflectable by the ionosphere.

*The radio horizon refers to the distance in which radio signals from two points are effectively blocked due to the curvature and curvature of the Earth which means that the Earth appears less curvilinear to radio waves than to light. This results in those UHF as well as VHF waves travel around 15% more than the actual distance of the line.

A multi-path distortion can occur when the receiver receives radio signals through two or more routes which means that the signal could reflect off objects. If it is reported by another operator that the signals from your station's 2 meters were powerful a while before, but have become weak or deformed, alter directions of the antenna, or move away a couple of feet to assist. Data transissions can cause errors. rates can increase, especially when VHF or UHF signaling are transmitted across multiple routes.

* Because of less absorption by plants The spectrum that is available for VHF as well as UHZ signals might increase in the winter months.
* A repeater is a transmitter and receiver which is likely situated in an elevated location that can extend the range of your radio by repeating the signal. If you are using a repeater you can also make use of a multi-path. This will benefit you, even if you don't have a direct line that is visible to the Repeater. If you're using an antenna that is directional, seek out a route which reflects signals into the repeater.
UHF radios can be better employed indoors as their shorter wavelength gives them more entry into the structure of buildings.

* Knife-Like Differfraction occurs where signals have a partial refracted by certain solid objects that have sharp edges, such as building.

The direction of an electric field, or antenna polarization with respect to VHF as well as UHF frequency is extremely important. Signals may be weaker if the receiver and transmitter don't use similar polarization. Horizontal polarization is commonly used to transmit weak long distance signals CW for example, like "morse code" and repeaters use vertical Polarization.

* Picket Fencing refers to a rapid flapping sound that is usually heard from a mobile device that transmits while moving.

* When signals from the 10,6 two meter band are detected over a long distance, the signal may be reflected by an sporadic layer of E of the Ionosphere. The sporadic E propagation is likely to be associated with strong over-the-horizon signals in the 2, and 10,6 meter bands.

Some other phenomena that can occur that occur at VHF frequencies include auroral reflection as well as the troposhereic scatter of meteors, and trosposperic ducting. If VHF messages are received through auroral reflection, the signals at most times change in strength, and sound quality is distorted.

The 6 meter band is the most suitable band to communicate through meteor scatter.

* Tropospheric scatter is the method that allows the over-the-horizon VHF and UHF communication to distances of around 300 miles. This is on frequent basis.

* Troposperic ducting is caused by temperatures that are inverted in the air and propagate VHF signals for hundreds of miles

* Light rain and fog generally have no effect on the 6 and 10 meter bands because prespiration is not a significant factor on lower frequency.

*Ionospere is a part of the atmosphere that allows the transmission of radio signals across the globe.

* Ionospheric propagation at Lond distance is mostly utilized on the HF band and is a benefit of high frequency radio when compared to VHF and other frequencies.

* At the height that the sunpot cycle that typically happens every eleven years, the 6 or 10 meters band could be used to provide long-distance communication as frequency of HF is regularly.

* Dawn until shortly after sunset, during times of intense sunpot activity. This is the most optimal time to get long-distance propagation of the 10meter band through the F layer.

• Unusual loss of signals from distant stations, despite having good reception is the result of multiple signalsthat travel through various routes.

Radio Services

If you do not have a Ham radio license or any other license that is commercially valid You will be restricted to a small number of relatively low-power radio communication within the United States, some which include:

* 49MHF - Very Low Power Walkie-Talkies with short range or cordless phones.

* 900MHF- Also a fairly low power cordless phone that has a short range.

* FRS* FRS Family Radio Service- 460MHz UHF, with a limit of 1/2 watts power, for a brief distance

* CB Radio- license free Limit of 4 watts, CB Radio can transmit data over numerous miles along the line of vision.

* MURS - Multiple Use Radio System, 150MHz VHF. It also connects several miles away

* GMRS* GMRS General Mobile Radio Service, it needs an FCC license, but provides more range and power than regular FRS however, it uses the same frequencies.

Chapter 4: The Electronic Basics

Electronics Concepts

* A Resistor is an electrical element that blocks the flow of electricity in an DC circuit. The majority of resistors have fixed values in ohms which is known as the resistance.

"A Variable Resistor" is an "potentiometer" that allows you to alter the volume of the device by varying the resistance.

A capacitor is an electronic component which has two or more surfaces that are conductive, separated by an insulation. It stores energy within the form of an electrical field. Capacitance is the capacity to store energy. It is expressed by "farad".

* An inductor stores energy within a magnetic field that is typically comprised of coil wire. Inductance refers to the ability to store energy within the magnetic field. It is measured using the unit known as "henry."

* A switch is an electric component that is utilized to connect to or disengage electrical circuits.

"Fuses" are electric device that protects various circuit elements from being overwhelmed

* A battery is a device used to store the electrical energy chemically

A diode is an electronic device that permits the flow of current only in only one direction. Diodes have two electrodes: an anode as well as an cathode. A stripe separates the cathode of a semiconductor diode.

* Light Emitting Diodes also known as LEDS These are a kind of diode that emits light when electricity is flowing through them. They are typically used to show a visual indication.

The term "transistor" refers to an electronic device that can use either a current or voltage signal to regulate the flow of current like amplifying a signal. "Gain" refers to the ability of a device to amplify an audio signal. A transistor is comprised up of three different layers made of semiconductor materials

* This component of the electrical circuit can be utilized as the primary gain producing element in the RF power amplifier

There are two kinds of transistors with bipolar junctions.

PNP- This is composed of two players and an N layer. This causes it to carry an overall net positive charge

NPN- This includes two N layers and one player, making it net negative charges

They are both composed of three electrodes, Base, Emitter and Collector.

* Field Event Transistor (FET) also contains 3 electrodes: Source, Drain and Gate

Understanding Schematics

A diagram of electrical wiring that employs standard components within its layout is known as"schematic" "schematic". A precise connection of electrical components can be illustrated in the schematic diagram of electrical circuits.

* Figure T1 is a diagram of a lamp that is turned to the right position when positive chaege placed on the source.

* Component 1 is a resistor which restricts the current input
* Component 2 is an electronic device that regulates the flow of current that goes the lamp. This will turn it on and off.
* Component 3 is the lamp.
* Component 4 is the battery.
* In figure T2 is a basic power supply 3. It is one pole, one throw switch that switches the power supply off and on.
* Component 4 is a transformer that converts 120V AC house current into lower AC voltage for various purposes,
* Component 5 can be described as a rectifier diode which converts AC into various DC signals
* Component 6 is a capacitor. this instance it's job is to block the 60Hz portion of the rectified AC
* Component 8 is an LED that lights up when the power source is turned on.
* Component 9 is a variable resistor. Its objective is to reduce and regulate the output current

* In figure T3, an output of the transmitter circuit.

In Figure 3 Component 3, a variable-inductor, it is used in conjunction with a capacitor in order to create an optimum circuit.

4. Component 4 represents an atenna

* A simple resonant , also known as tuned circuit is an inductor and capacitor that is connected either in parallel or series to create the filter

Series Circuit

An example of a series circuit would be one where the current is identical across all components.

* The current remains constant at the junction of two components within an in series circuit

* The voltage of each of two components connected to the voltage source is determined by the type and the value of the components

Parallel Circuit

It is form of voltage where the voltage is identical across all components

* The voltage across two components that are in parallel to an electric source is the same as the source voltage. the source
* The current that flows through two components in parallel is split in two parts, according to the component's value

Other Electronic Components

A ralay is an electronic stwitch that is controlled by electricity.

* A meter can be used to show an electrical quantity in an n-digit value

"Integrated circuits" is a reference to a circuit which combines various semiconductors and other components into a single package. ICs can be used for digital or analog functions.

* A typical analog Integrated circuit is the voltage regulator. It is a kind of circuit that regulates the quantity of voltage that is generated by an energy source.

* When connecting circuits we employ conductors that have shields around them, that is connected to ground. This is done to stop the transfer of undesirable signals to or out of the wire.

Chapter 5: Antennas And Feed-Lines

Antennas and Feed-Lines

An antenna that is horizontally polarized will be perpendicular to the Earth The vertical one is parallel to Earth

A half-wave dipole is an antenna measuring roughly one half wavelength from the end-to-end. The direction from which radiation is most intense from an antenna with a half-wave dipole in an open space is broadside of the antenna

* Due to the stray capacitance of objects that are close in proximity to an antenna such as the ground, a typical dipole in reality 5percent shorter than half the wavelength. For example, the approximate length of a six-meter half-wave antenna is 112"(inches). In order for this dipole to resonate at a higher frequency, you have to reduce its length.

The typical dipole is horizontally poralized. A commonly used vertically polarized antenna can be that of the vertical quarter wave .

* The approximate length of a vertical antenna for the frequency 146MHz is 19"(inches)

A loading coil can be used to make an antenna shorter by inserting an inductor into the radiating area of the antenna to make it electriaclly longer.

The beam antenna,, also called an directional antenna is one that concentrates signals in a single direction. Certain directional antennas are the quad Yagi or dish antennas.

An antenna's gain is defined as the improvement in the strength of signal in a specific direction in comparison to the antenna used as a reference.

In the majority of cases hand-held radios employ"rubber duck" antennas "rubber duck" antenna. It doesn't communicate or transmit as efficiently as the large-sized antenna. If this kind of antenna is utilized in a vehicle, signals might not be clear due to the shielding effects on the automobile.

* Mobile UHF and VHF antennas are typically placed in the middle of the roof of the vehicle, since roof-mounted antennas provide the most homogeneous radiation pattern.

An appropriately mounted 5/8 wave VHF or UHF antenna will typically have less radiation angle and better gain than a an antenna of 1/4 wavelength.

Feedlines serve in order to link your radio to an antenna. The majority of the time coaxial cables are utilized to complete this job, since it is simple to use and requires some special setup procedures.

* As the signal that passes via the cable grows the loss will increase.

When selecting an antenna, it's crucial to select an antenna that is of the same impedance as the one being used.

*Impedance is the quantity of resistance of current against the AC current flowing through the circuit. Ohms are the units used to measure impedance.

The majority of Amateur radios come with output impedances of up to 50 Ohms The most widely utilized coax cable used in Amateur radios is 50 Ohms. examples: RG - 58 and RG - 8

* RG-8 cable which is the biggest, has lower loss at a certain frequency

* Air-insulated lines have the least percentage of loss when used at VHF and UHF frequencies.

The most common cause of the failure of coax cables can be ""moisture contamination"

* Ultraviolet light may damage the jacket and lead to water entering the cable.

One drawback of the coax cable made of air when contrasted with the solid or dielectric type of foam it that it requires special knowledge to stop water absorption

* The coax cable connectors PL-259 are the most frequently used for HF frequencies.

The Type N connector is the most to be used for frequencies higher than 400MHz.

* To stop an increase in the loss of feedline all cables must be sealed well against water intrusion

An irregular change in SWR readings could an untight connection in the feedline or antenna, ensure that your connections are in good order.

* The Standing Wave ratio a measure of how well an antenna is aligned to the transmit line. This is when they share the identical impedance.

It is essential to be able to achieve an extremely low SWR so as to enable efficient transmission of energy, and decreases signal losses.

* SWR is measured by using an SWR meter. It should be connected in series to the feedline, and between the transmitter and the antenna. It must be close to the transmitter.

A directional wattmeter could be used to determine SWR by measuring the power reflected and forward and then formulating the SWR.

* The majority of solid state amateur radio transmitters decrease their output power when SWR grows to protect amplifiers that are used for output.

An antenna turner could serve as a tool to align the antenna's amplified to that of the transceiver's output impedance.

A dummy load can be utilized to avoid transmitting signals through the air when testing is running and it is comprised of a resistor that is not inductive and a heat sink.

* To make sure that the antenna's frequency is the frequency you want to operate at Use an antenna analyzer.

A turner, or an antenna coupler matches the antenna's's impedance with the output of the transceiver.

Chapter 6: Modulation And Signals

Modulation and Signals

"Frequency Modulation" (FM) is the most commonly used modulation to operate VHF Voice repeaters UHF and VHF as for VHF satellite radio. FM alters the frequencies of the waves to transmit data. It cannot be used for transmitting HF frequencies due to the fact of the bandwidth.

AM, also known as Amplitude Modulation (AM) is a method of varying the amplitude, or in other cases the strength of the signal used to transmit information. AM is radio's oldest transmission

* Single Sideband (SSB) This is a variant of AM. However, in this instance the carrier is also one of its sidebands are is removed. SSB Voice modulation can be mainly utilized for long-distance signal contacts with VHF and UHF as well as bands for HF. It can also be utilized for an upper-sideband (USB),or as an lower sideband (LSB)

The upper sideband is typically used for 10 meters UHF and HF VHF single-side-band communications

The main benefit of the single sideband signal over FM transmissions lies in the fact it is that SSB signals have a smaller bandwidth, which is 3kHz in essence in comparison to the VHF repeater signal for FM phones, which is between 10 to 15kHz.

* SSB phones are employed in all bands that exceed 50MHz.

*The Morse code also referred to as "continuous wave" is the kind of emission that has the smallest bandwidth, having only 150Hz

* If you are sending continuously wave (CW) using amateur band, International morse code is used. code used.

If you are transmitting CW it is possible to utilize straight keys or an electronic keyer, or an electronic keyboard

* Analog fast-scan TV broadcasts in the 70cm spectrum are just a few of the Modes with huge bandwidths, approximately 6 MHz.

* The kind of transmission referred to by the word NTSC refers to an analog color TV signal

Digital modes are able to transmit digital data instead of analog or voice signals. Examples of digital modes are: packet radio as well as IEEE 802.11 and JT65

A radio signal contains the check sum, which allows error detection and detection. This is the header that has the call number of the station which the data is transmitted, and then, the station will automatically repeat the request in the event of error.

The Automatic Packet Reporting System (APRS) uses the packet radio as well as the Global Positioning System (GPS) receiver for sending automated location reports using the amateur radio. It offers real-time, tactical digital communications, and displays maps of stations' locations.

*The phase Shift Keying (PKS) transmits data at a low rate so it is known as a low rate mode for data transmission

* A ARQ transmission system transmits a digital code, the receiver station detects an error and sends a message to the sending station to send the data

* Digital Mobile Radio(DMR), is a method of multi-plexing digital signals in the same 12. 5KHz repeater channel

"A DMR "talk group" is a method that groups can use to share the same channel at various times without any other listeners hearing them within the channel. The group's ID, or Code is used to program your radio when you decide to join the repeater "talk group"

* WSJT Software provides weak digital communications modes for Amateur radio channels. A singleband transceiver with a personal computer equipped with an audio card is required in order to communicate with any or more of the modes WSJT offers. WSJT suite provides access to a variety of activities like Earth-moon Earth or Moonbounce beacons that propagate weak signals as well as meteor scatter.

* The FT8 digital mode capable of operating under low signal-to noise environment. It can transmit at intervals of 15 seconds.

* High-speed multi-media network is also known in the form of Broadband Hamnet (TM), is an amateur-radio-based data network that uses commercial Wi-Fi equipment with modified firmware

Chapter 7: Tips For Safety

Electrical Safety

* Safety must be taken with care, as the lethal voltage could reach as low as 30V, and the lethal current could be at as low as 100mA which isn't that excessive.

* Allowing current to flow through the body may result in health problems, disrupting cell's electrical functions, and may even trigger involuntary muscle contractions

* The safety ground wire is connected with the green wire on the three wiring electrical AC plugs, and is the one to use for amateur equipment too. The risk of electrical shock at your workstation can be reduced by following these steps:

Make use of three-wire cords as well as plugs to connect your AC powered equipment

All AC Station equipment have to be connected with a standard ground for safety

Use a circuit protected by a ground fault interrupter

* If you are in requirement to repair a fuse ensure that you replace the fuse that has been blown with the same type and amount. It is a mistake to replace a fuse with a 20A rating with a 5A fuse because excessive current could cause an outbreak of fire.

* A fuse circuit-breaker that is in series with an AC "hot" conductors should be built into all home-built equipment that is powered by 120V AC circuits

When working on device, shut off your power supply even though, in other instances, you may get an electric shock from stored charge in the case of capacitors with large capacities. Be cautious when working with capacitors.

When it comes to batteries, shorting their terminals could cause fire, and even explosives. If a lead-acid storage device is discharged or charged too fast the battery can overheat and explode.

* When you are putting up your antenna towers, ensure that you're not in the vicinity of electrical wires that have been damaged, and place your antennas away from power lines in order to be secure in the event of an sudden falls of antennas. It will not be in close proximity to power lines. Additionally, keep the antennas from the poles that are used for ultility since they can come in contact with high voltage wires for power.

* The antenna must be placed in an arrangement that no one is able to touch it when you transmit in case this results in an uncomfortable RF burn.

* The loosening man line due to vibrationcan be avoided by using a safety line via an buckle.

A gin pole must be used in the construction of the tower, in order to raise tower sections or antennas.

* A hard hat and safety glasses are required for every member of the tower's team of workers any time there is work that needs to be carried out in the tower.

Don't climb a building without assistance and an observer.

* Put on a carefully verified climbing harness before climbing on a ladder; wear a fall stopper and safety glass.

Grounding is crucial due to the fact that towers are basically lightning rod. Additionally, local electrical codes set grounding specifications for amateur radios or antennas. Conductive straps or wires should be used to join earth connections or external ground rods.

Beware of abrupt bends while connecting conductors to ground. For lightning protection. Connections must be direct and short.

* To ensure that a tower is properly installed, you must place separately 8-foot rods for ground, as they pertain to each leg of the tower and bonded to the tower along with the rods.

* The VHF or UHF radio signal are not ionizing radiations. They are as distinct from radio waves that are ionizing because RF radiation is not able to generate the

power to cause genetic harm. Be aware that even with this the radiation is still risky as the highest power level that an amteur radio station can utilize at VHF frequencies prior to when an evaluation of RF is needed is 50 watts PEP(peak envelope power) at the antenna.

To determine the frequency of your radio exposure to verify if your station is compliant with FCC regulations, the following steps are possible to follow:

Calculate using the FCC OET bulletin 65

Calculate based on computer-generated models

Assess the strength of the area with equipment calibrated

* The proportion of time the transmitter is transmitting is known as "Duty cycle" It can be one of the main factors to determine the safe exposure to radiation levels, since it impacts the amount of exposure that people are exposed to radiation.

* If the typical time to expose is 6 minutes two times as much power density is allowed when the signal is on for 3 minutes and then off for 3minutes instead of being present throughout the duration of 6 minutes.

* Limits for exposure vary with frequency because the human body is more able to absorb frequencies. RF energy is absorbed at a certain level of frequencies than other. The 50MHz band is the lowest maximum as well as the lowest exposure limit that is permissible. The following are the factors that affect exposure to radio frequency radiation;

The power and frequency in the field of RF

The space between an antenna and the person

The pattern of radiation from the antenna

* To cut down on radiofrequency radiation that is beyond the limits set by the FCC, amateurs might think about relocating their antenna

After your initial RF exposure test, make sure your station remains in conformity with RF safety standards by reviewing your station every time there is a change in the equipment.

Chapter 8: Station Set-Up

The Set-up Your Station

When making your station ready, these points should be taken into account:

* Make sure that the wires connecting the power source as well as the radio station is a high-quality wire. It should be as thin as you can in order to prevent voltage from falling.

* To determine the capacity minimum required by a transceiver, think about the following:

- Effectiveness of the transmitter when it is at the maximum power output

Control circuit power and receiver circuit power

- Controlling the power supply and dissipation of heat

* Computers can be employed in the amateur radio station

Contacts can be logged in to access contacts and other information

- To transmit or receive constant waves

To generate and decode digital signals to receive audio, send audio and even push-to-talk connections can be utilized in conjunction with a transceiver for voice as well as the computer to perform a digital operation

The microphne in the computer's sound card may be connected to the headphone of a transceiver for digital mode operation.

A computer sound card, you can provide audio for the microphone. It also transforms the audio it receives into digital, in the event that you are transmit a digital communications

A ferrite choke utilized to limit the amount of RF moving through the protective shield on an power or audio cable, it can also correct distorted audio

The flat strap can be described as the most suitable conductor employed to provide RF grounding within your station because it is the most low impedance signals to RF

If the battery is connected to the engine block's ground strap the mobile transceiver's power negative connection is required.

The alternator produces an extremely high-pitched whine that is distinct from the engine's speed in a mobile transceiver's receiver audio. However, filters can be employed to reduce the effect of this. Noise blankers eliminate the engine noise or power line However, one way to minimize the interference from ignition for the receiver is to turn off the blanker.

If a transmitter is controlled by setting the microphone's gain to too much, it alters the output signal

If you record the frequency inside memory channels that allows quick access to your preferred frequency on your transceiver

The squelch switch on the transceiver reduces the output noise of the receiver when a the signal is received. The squelch must not be set too high , as it can block all signals.

The term "reciver offset" refers to the difference in the repeater's transmit and receiver frequencies.

"RIT" - "RIT" is a reference to the receiver's incremental turning. The receiver RIT is used to adjust the pitch of a single side band signal, that appears to be to be too high or low.

- To ensure that the audio received is at a constant and steady level using an Automati Gain Control used. employed.

The benefit of having a multimode transceiver having different bandwidths for receiving is that it allows noise reduction through choosing the appropriate bandwidth to match the mode.

A device that combines both functions of receiver and a transmitter the receiver is referred to as"transceiver "transceiver"

A transverter converts input and output of a tranceiver into another band.

In order to increase efficiency of the handheld transmitter an RF power source can be employed.

Selectivity and sensitivity are the two main specifications of the receiver. The phrase "sensitivity" means the receiver's capability to detect signals. While "selectivity" means the receiver's ability to differentiate between various signals. A RF preamplifier may be utilized to increase the receiver's sensitiveness. For optimal performance, it should be installed in between your antenna and receiver.

- In order to convert radio signals from one frequency to another one, it is called a "mixer" is utilized.

The oscillator can be described as the device that creates the signal to achieve a certain frequency

"Modulation" - The term "modulation" refers to the combining of speech signals with the RF carrier signal

Radio Frequency Interference

Radio frequency interference may be the result of one or more of these

- Some spurious emissions
- Fundamental overload

Harmonics

If someone complains about the transmission of your station interfering with his TV or radio signals, first ensure that your station is operating correctly and is not interfering even with your TV or radio frequency in the event that you are tuned to the same channel.

The first step to eliminate a cabble, or TV signal from your radio amateur is to check that the coaxial connectors are put in place. If the receiver is not able to stop a signal that is strong that are not in an AM or FM bands it could result in unintentional transmissions by other FM or AM radio bands.

The band-reject filter may be used to minimize the load on the VHF transceiver by the nearby FM station

- To cut down on the interference caused by an amateur radio transmitter with the phone in close proximity, install an RF filter in the phone.

To minimize interferance issues You can think about the following devices:

- Snap-on ferrit chokes

Band pass and filtering for rejects

High-pass and low-pass filters

It is possible that people around you possess a wireless device commonly referred to Part 15 devices, these devices are not licensed and can emit low-powered radio signals at frequencies utilized by licensed services which could disrupt your station If a device at the house of your neighbor is interfering with your station

Be patient and identify the specific device by your neighbor

Talk to them politely or her, making them know the guidelines that prohibit him from using such devices

Make sure to check your station to make sure that it's in compliance with the best practices of amateur radio.

* If feedback is received from you, regarding the audio quality of your radio it could be due to an outcome of

Your transmitter is slightly off the frequency

If your batteries are being low

- You being in a bad location

Reports of distorted voice transmission could be due to an RF feedback within the transmitter.

If your radio amateur transceiver is causing too much deviation you, try speaking further away of the transmitter.

Chapter 9: Operating Your Amateur Station

Check out your station's performance

The voltage and the resistance are two common measurements taken with the multimeter.

A voltmeter is utilized to measure force generated by electronic devices

The voltmeter is connected an analog circuit to the circuit

If you are taking measurements of high voltages using the voltmeter, ensure that the voltmeter and its leads are designed for use at the voltages being determined.

A Ohmmeter is a device that measures resistance. You shouldn't use it to measure circuit resistance while it's powered.

If a circuit has the largest capacitor, and an Ohmmeter is attached to the circuit, at first it will display an extremely low resistance. However, the resistance will rise with time, due to the capacitor's large size.

Ammeter is an instrument that's used to measure electrical current. Usually is connected to a circuit in line with the circuit.

As an amateur user it is crucial to master the art of soldering. To get the best results, Rosin-core solder is suggested for electronic and radio work. Surfaces of solder-joint that is cold is usually "dull"

Operating your Amateur Station

Repeater stations retransmit the signal from another amateur station using the same channel. A repeaterstation is able to automatically retransmit signals of other stations.

A linked repeaters network sends signals from one repeater and after that, it repeats by the rest of the repeaters.

Repeaters receive signals at one frequency, and transmit them on another frequency.

The difference between the frequency of transmission and the frequency of reception is referred to as "repeater frequency offset". It is possible to program your radio so that it receives signals using the frequency of the repeater's transmit and also transmit signals on the frequency that the repeater receives. The most commonly used frequency of repeaters offset in the 2m band is: and minus, 600KHz in the 70cm range plus or minus 5MHz, which is the standard offset frequency of repeaters.

A station that is amateur and sending and receiving at the same frequency can be described as an "simplex communication".

- When an amateur radio station is operating using "repeaters" It is called duplex operation.

Amateurs using simplex will be able to find all frequencies in each band have been designated to be used as "national call frequencies". The national call frequency for FM simplex operation on 70 cm band 446.000MHz in the 2m, it's 146.52MHz. These frequencies are offered in order to allow stations that are within a mutual communication range to connect without the need for a repeater.

Repeaters are programmed to operate in areas with high levels of interference. They are not able to function unless the station they receive from transmits the subaudible frequency of specific frequency. CTCSS is the usage of a sub-audible tone. It transmits using normal voice to enable the squelch of the receiver. If you don't have this feature for your radio your repeater will not be able repeat the transmission.

If the output of a repeater is being heard, but it's not easy to get it into your system It could be due to the following reasons: the reason:

Incorrect offset of the transmitter

The repeater may require a CTCSS tone that is appropriate from the transceiver

The repeater could require an appropriate DCS tone from the transceiver

If the squelch of the repeater remains open, keeping an eye on the frequency of the input to the repeater could allow you to hear the station calling the repeater, in case your station's signal isn't strong enough.

If a repeater's contact informs you that your transmissions are breaking up at voice peaks This means you're talking too loudly.

When you use a repeater it is mandatory to say your call number; KB9OKB listening. This will indicate that you are paying attention

If you want to call someone who is on a repeater you need to mention the station's call number and then your name; KB9OPV, KB9OKB.

If you wish to connect with another station on the HF band by calling "CQ" You could use something like "CQ,CQ,CQ This is KB0OKB". CQ calls any station

- Prior to deciding on the frequency at which to call CQ,
- Ensure that no other person uses the frequency.

Inquire about whether the station is currently in use

Make sure that you're in the band you have been assigned to.

When you respond to a CQ from a call begin transmitting using the call sign of the station, prior to your own call sign. When a test transmission is activated, identification of the station must be used at least once every ten minutes during the test as well as at the conclusion.

Every technician should be able to use Morse Code on specific portions of the 80m, 40, 15, as well as the 10m band. Amateurs typically employ three-letter combinations known as Q-Signals. The QRM signal is one of them. QRM signals are the "Q" signal that informs you that you're getting interference from different stations and you will receive the "Q" signal which informs you that you are changing frequency is called "QSY".

While it is advisable to review a band plan prior to operating it is important to remember that the band plan is an unofficial guildeline to use various modes of operation within an amateur.

When using your radio in normal conditions and not in distress the power limits advised by the FCC states that you should not overpower the maximum power that is permitted in a specific band. just the minimum amount of power required for the particular communications.

If two stations operating at the same frequency clash the common sense should prevail as no one individual can have any claim to the frequency of an amateur.

When you identify your station by phone or a phone, a phonetic alphabet should be employed by the FCC for example, those of the NATO alphabets that are phonetic.

• RACES' Radio Amateur Civil Emergency Service and ARES; Amateur Radio Emergency Service are two organizations providing emergency communications. Both have an important feature that is shared by both, can provide communication in emergencies. The group consisting of amateurs licensed that voluntarily sign up their credentials and equipment used for communications within the public service are known as "ARES".

The term RACES means;

A radio service that operates on amateur frequencies, to manage civil defense or emergency communications

A radio service that is based via amateur stations to handle civil defense communications in emergencies

A crisis service that employs amateur operators who is certified through the Civil Defense agency as members of the organization.

"Net" - "net" is an internet network that is usually created from amateur radio radios in the event that emergencies arise to improve communication. It is managed through the NCS; Net Control Station which is responsible to transmit messages promptly and effectively.

Amateur operators who are not the control station for the net are known to "check into the net". After an amateur has checked in to an emergency traffic network the operator is not able to operate on another frequency. Instead, he stays on the frequency until otherwise specified by the station controlling the traffic net.

To receive immediate response from the net controller, an amateur must begin his call with the words "priority or urgent" and then follow his signal

A reliable emergency traffic control system is distinguished by its capacity to transmit messages once it has been received. The message traffic can be informal or formal

A standard phonetic alphabet is used to make sure that voice message messages that contain names as well as some odd words are correctly copied by the receiver station.

A formal traffic announcement is composed of four elements: the preamble address, the preamble, the text, and finally the signature

The Preamble in the formal traffic message is the data needed to keep track of the message when it goes through the amateur traffic control station. "Check" is an element of the Preamble and is an indication of the number of words or words within the text part of the message

The Address includes the address and the address of the person who will be receiving it.

The Text the message to be sent

The Signature - The Signature - Identifies the aurthor , or the origin of the message

Station operators operating on amateur radio are allowed to operate beyond the frequencies of their license, as per FCC regulations, but only in situations that require the immediate security of life and property.

Some of these are employed in amateur radio satellites.

- SSB

FM

- CW/Data

To stop access to others, an effective radiation power source on an uplink to satellite is needed

To determine whether your upright power is correct, your signal strength from your downlink should be close to the strength of the beacon.

Most of the time it is the case that the uplink and downlink frequencies fall in various amateur band. For example, if the satellite is running within "mode U/V" that means it is operating in the 70cm (UHF) band, and the downlink is located in the 2 meter (VHF) frequency.

Amateur operators who holds a Technician or Higher Class License is able to communicate with an amteur station on the international space with the 70cm and 2meter bands amateur frequencies.

- LEO is used to define the satellite that operates in the Low Earth Orbit

A status beacon is the signal from a stallite, which includes details about the health and state of the rock.

The most common method of getting and sending signals from digital satellites is through the FM packet

Anyone who can be connected to telemetry signals will be able to get the telemetry of an orbiting stallite

A stallite tracking software can be used to identify the space in which a space stallite from amateurs can be located and can be used to provide

Maps that show the actual location on the trace of the stallite on the globe is shown.

It displays the time as well as azimuth, elevation at the beginning point, the highest altitude and also the time at which a pass is completed.

The actual frequency of the transmission as well as signal that shows the effects of Doppler shift

The inputs to the stallite tracking software are called the "keplerian components".

Doppler shift and Spin fade are two issues you need to address when communicating with stallite. Doppler shift refers to the variation in the frequency of the signal due to relative motion between the stallite's antenna and the earth station and the constant movement of the stallite and its antennas results in slow fading of the stallite signals.

Contesting is an operational activity that reaches out to as many stations as is possible within a specified time

Radio Direction Finding This involves using radio direction-finding equipments and Sklls in order to participate in a transmitter hunts. It is also known as"fox hunt" "fox hunt" an antenna for direction is required in the course of this. Radio direction finding can also be an instrument used to find the causes of interference from noise or jamming during transmission.

The maximum amount of power permitted when transmitting telecommand signals radio controlled models is 1W

To link amateur radio networks like repeaters to the internet via the voice-over internet protocol the technique is known as "the ineternet radio linking Project, (IRLP)", is used.

When the station is connected to either the Echolink or IRLP it is referred to as a node. A IRLP node is accessible making use of DTMF signals. To select a particular node of IRLP, when transmitting via a portable transceiver keyboard is required.

The call sign you choose to use must be registered, and an evidence of license is required prior to being eligible to Echolink system to talk to repeaters.

- To connect to the active nodes with VOIP; Subscribe to an online service

A list of on-line repeaters kept by the local frequency coordinator for repeaters
-Use a repeater directory

A gateway can be described as an radio amateur station which connects amateur radio stations to the internet.

Chapter 10: Fcc Rules And Guidelines

FCC Rules of the FCC

The radio amateur service is managed through the Federal Communocation Commission (FCC) The FCC regulates and enforces rules for the amateur radio services in the United States. Part 97, which is the portion of the FCC regulations that governs to the rules for the radio amateur service.

* The principal goal of radio amateurs is to increase the knowledge and skills of the communication and technical aspects of radio science which is stated within the FCC rules and rules.

*In Part 97, a beacon is defined as an amateur radio station that transmits with the intention of observing propagation and other similar research.

* The Part 97 defines space stations as amateur station that is located at least 50 km above the surface of the Earth.

* The intentional interference with radio stations is not at any time allowed, as per the FCC rules.

* The transmit/receive channel as well as other parameters for repeaters and auxillary stations must be suggested by an "volunteer frequency coordinator" and is accepted by local amateurs.

* Amateur operators within a specific regions, whose stations can qualify to be repeater stations, are under the responsibility of selecting or choosing the frequency coordinator

* ITU * ITU; International Telecommunications Union, is the authority in charge of information and communications issues for the United Nations

* If you are transmitting at 146.52MHz that is in the 2 meter band

* You should not set your transmit frequency to the fringe of an sub-band or band for amateurs This is for:

* Allow for error in calibration in the display of frequency for the transmitter.

Sidebands for control modulation with a band edge stretching beyond the edge of the band

* To account for transmitter frequency drift

* The permitted limit for emissions in the frequency range of 219 and 220MHz is only recommended for fixed digital message forwarding systems

*10 Meter band is the HF band where Technician class operators are granted the privilege of using RTTY or phone.

* The following frequency ranges are only limited to the CW 50MHz to 50.1MHz in addition to 144.0 to 144.1MHz

* The maximum acceptable peak envelope power output for Technician class operators operating on the designated bands of the HF spectrum will be "200watts".

* The maximum amount of power for Technician class operators operating at frequencies that exceed 30MHz is 1500watts, unless specifically stated in the limitations.

* So long as amateur radio is only available in certain parts of amateur bands, US amateurs could encounter non-amateur stations within portions like 70cm band. they should not interfere with them.

* The typical term used for an FCC issued primary station, or an the operator license granted to amateur radio is 10 years.

Once your name and name appear within the database of FCC's ULS database, once you've completed and completed the exam required to obtain your first license as an amateur, you're able for operation of a radio transmitter using the Amateur Service Frequency.

* A person can't have more than one primary or operator station license.

* The control operator's primary station license has to be displayed in the FCC ULS central license database as document proving his possession of the FCC issued operator/primary license

* Any licensed amateur may select any call sign one wishes within the rule of vanity call signs.

* For each FCC-issued primary or operator license the call sign for that station is composed in one of two or more letters which is followed closely by a numerical number and then a letter. three letters, as the case may be. K1CCC is an example an acceptable call sign.

* If an amateur license is due to expire, a two-year grace period is granted within which the license may be renewed.

* A license for amateur radio that has expired is not valid for transmitting as long as the FCC license database confirms that the license was renewed even though it's still within grace period.

* In order for a club to receive an "club station permit grant" The club must be at least four members.

When a grantee is unable to supply and maintain the correct address for mailing with the FCC it could result in the suspension or revocation the operator's license or station license.

* Unless a country has a favorable opinion It is a crime to operate an amateur radio station in a different country.

* An FCC licensed amateur station is permitted to communicate from any boat operating in the international waters, and is documented and registered with the United States

The use of vulgar and offensive language is restricted in radio stations that broadcast amateur radio

* The transmission of music from an amateur radio with a phone transmission is only permitted in the context of an authorized re-transmission in a spacecraft manned by a human .

It is not permitted to transmit encrypted messages to obscure their meanings while transmitting commands for control to a space vessel or radio controlled craft.

* A FCC-licensed station cannot exchange messages with any nation whose administration has informed the ITU that it is opposed to these communications, even

though there are none at the present moment.

* Radio operators who are amateurs are not allowed to make money , unless in exceptional circumstances, for instance, if the broadcast is in connection with teaching in a classroom in an educational institution. the radio operator could be paid a fee.

* When equipment that are used in amateur radio operations are available for sale, an amateur radio operators can use their stations to inform other amateurs about the availability of the equipment available for purchase, but it is not frequently carried out.

Amateur operators are not allowed to broadcast. Their communications must be directed to a station. The phrase "broadcasting" in the FCC guidelines for the amateur service is broadcasts that are designed to be viewed by the general public.

* An amteur operator can only broadcast when transmitting code practice

information bulletins, transmissions or code practice essential for emergency communications.

Amateur stations are authorized to transmit broadcasting signals or news gathering, as well as production of programs, but only when this is the only method for transmitting such information, and when it is could be life-threatening with regards to property and life.

* An amateur operator can only transmit one-way in the case of transmitting codes or information bulletins, or transmitting messages essential to get emergency communications

* International communications permitted to an FCC licensed amateur station are communications that are incidental to the purposes of amateur radio and emarks with personal significance.

* The person who holds the license for the station is assumed to be the FCC to be the control station's operator. station unless stated otherwise in the station's logs.

* Amateur stations cannot transmit without a controller operator. If the licensee of the station is not the control operator of the station and he is not the station control operator, he has to select one who he is in close contact with for the smooth running for the station. If a repeater is inadvertently transmitting any communication that is in violation of FCC guidelines, the source station is held responsible.

* The privileges granted to an amateur station varies on the type of operator license that is held by the controller. Technicians with a class license cannot be a controller, operating in the extra class operator segment within Amateur bands.

If you are an operator who is amateur in the event that your license allows you to broadcast on the satellite's uplink, you could also be the controller of a station that is communicating via an amateur satellite.

The term "control type" and the control point are closely-related terms. The term amteur refers to the place where the operation of the operator takes place. The type of control employed when transmitting using a hand-held radio is referred to as"local control. "local controller". Remote control, as defined in Part 97, operates the station via the internet. "Repeater "repeater" can be a good example for an automated control. The following is true about the remote control process:

The control operator should be present at his control station.

There must be a person in control always on the watch.

Invariably the operator of the control handles the controls.

* A station that is amateur-grade is expected to broadcast its call number at least every 10 minutes, throughout and after the communication.An amateur station is able to transmit without identifyingitself, in the case of transmitting signals to control an aircraft model.

* English Language is the only one that can be used to be used in the identification of stations particularly for those operating within a phone sub region. A phone signal transmitting station will transmit the call sign by using CW or phone Emission in order to provide the identification of the call sign.

* A self-assigned call signal marker such as"/3" can be utilized when operating a portable or mobile or to mark something regarding your location, such as KL7CCstroke W3

* Messages sent by the control operator to another amateur radio station's control operator in which the message originates from a different person is termed an inter-party communication. For example, if your guest comes to your home and you let them speak on your radio, this is considered a third party communication. Although it is acceptable for the United States, there are some restrictions when you need to connect with an amateur station that is not within the country. If you're not an amateur radio operator licensed by the government or radio operator, you may talk to an individual from a foreign country by using an unsupervised station of an Operator Class Technician provided that the country of origin has a third-party contract with that of the United States.

*The FCC and its officials must have an access right to the station as well as any other records that are related to it, at any time they wish or to conduct an inspection.

Chapter 11: Electrical And Electronic Parts That Make Up The Amateur Radio Station

A radio station for amateurs is an entire set of equipment and accessories for obtaining the required results. The station will be situated in an area where it can be used in times of emergency or in a disaster. The items mentioned in this article were chosen due to their compatibility with these requirements. Furthermore, you are able to utilize the same parts to create amateur radio stations when traveling with a handheld radio.

Flexible and Fixed Resistors

Fixed Resistors

They are made up of two kinds of metal oxide and carbon. The carbon variety is typically employed in consumer electronics due to its higher resistance to heat than the metal oxide.

Variable Resistors

They are employed in circuits that have to be controlled to change the amount of current or voltage that flows through them.

Based on the purpose they serve they are available in various shapes and sizes, but are generally classified into two major groups that are: potentiometers (pots) and rheostats (variable resistors).

Semiconductors

Semiconductors are substances that exhibit electrical conductivity that is shared between insulators and conductors. They are utilized in electronic circuits, such as diodes, transistorsand rectifiers as well as integrated circuits.

Circuit Diagrams

Circuit diagrams are visual representation of electrical components that make up circuits. They are used to create as well as to build as well as troubleshoot circuits in electronic devices. Circuit diagrams can also be employed to help teach electronic circuits and to design electronic devices.

Batteries

Batteries are electric device that stores energy through an chemical reaction. They are able for power supply to different devices, such as flashlights as well as portable radios. Batteries can also serve in backup sources of power for devices for medical use, lighting in emergencies and electric vehicles.

Rectifiers

An electronic device converts the alternating present (AC) into direct current (DC). This is referred to as rectifification. A rectifier alters how electricity flows moving it from one direction, to moving in the other direction. This makes it easier to regulate the amount of energy is used by a device within DC circuits.

The ability to power in various amounts can be beneficial when designing circuits that utilize motors, or other electrical devices , such as lights or speakers that require certain amounts of power.

Relays

Relays are switches that operate electrically. They can be used to regulate high-current circuits, or circuits that do not belong to an electrical circuit with the device being controlled. For instance, relays could be used to change the position of heavy-duty equipment like generators and motors.

They can also be used for switching lines among different electrical systems when one requires isolation from the other while they are working on a particular project. They have an internal coil which creates an electromagnet that draws pins in the relay to each other whenever power flows to the coil. After this occurs the coil lets voltage flow through it , and then makes contact with components of the circuit board!

The most popular kind of relay is known as"NC," which stands for "NC" or normally closed relay since there is no voltage applied to its terminals it doesn't do anything (it does not touch).

Voltage Regulators

Voltage regulators provide a constant voltage the load.

Voltage regulators are utilized in battery chargers, power supplies and DC-DC converters. They can transform a range of input voltages into the output voltage of a particular.

Meters

Meters are gadget that takes a measurement of a physical property and then displays the measurement in a format that is readable. The kind of parameter monitored determines the kind of meter that is used. For instance, if you need measured current you'll require an ammeter (also known as milliammeter). If you are trying to measure power or voltage it is recommended to utilize a suitable voltage meter or wattmeter.

"meter" or "meter" is used to refer to various types of equipment. These include:

* A device for measuring electrical quantities like voltage and current for a relatively affordable costs (e.g. VOMs)

* A device that is used to measure audio signals like frequency response or amplitude

* A device specifically designed to measure radiation (e.g., Geiger counter)

Indicators

Indicators display details to the users. Indicators could be basic devices, like bulbs for lighting or more sophisticated devices, like LCDs, or liquid crystal displays (LCDs).

Indicators are utilized in a variety of electronic devices.

Circuits that are integrated Circuits

Integrated circuits (ICs) are electronic miniature circuits made up of thousands or hundreds of tiny parts. They are used in many different applications, ranging from mobile phones to computers to toys and calculators.

The first ICs were invented in the 1950s by Jack St. Clair Kilby and Robert Noyce at Texas Instruments (TI). In the 1950s the semiconductor technology was quite new and transistors had been in use for a short time. However, TI was looking to develop smaller versions of components such as transistors that could then be used in the creation of new products such as televisions or radios. They believed this would give them more control over the way they put together electronics in one unit. And they were right!

A single IC could have hundreds or even thousands of tiny parts such as capacitors, resistors, diodes, inductors... and the list is endless! The size of the IC makes it ideal to be used in devices with small dimensions like cell phones and calculators since there's little room for error when creating something this small!

Transformers

Transformers are gadgets that transmit electricity between circuits the next through a change in the voltage. There are two kinds of transformers:

1. Electromagnetic, as well as
2. Electrostatic.

The former is employed in high-power transmitters, whereas the latter is utilized in receivers with low power. Both are made up of two coils twisted around a core of ferrite or iron materials (ferrite materials are very permeabil). The first coil (coil #1) is a lot of turns, and a comparatively small wire diameter in comparison to its second coil (#2) which has fewer turns and a larger wire diameter.

Resonant Circuit

The Resonant Circuit is a circuit that operates at an oscillation frequency that is natural. It is also an electronic circuit that is oscillating at its natural frequencyand thus, it doesn't require any other source of energy to keep this oscillation. Resonant circuits can be found in a variety of electronic devices, including radio transmitters, receivers television models, microwave ovens as well as radar system.

Resonance is when there's an inductive reactance that is greater than capacitive reactance , so it is not a net imperceptible (i.e. that"X L" = X C). In this case the resistance to load is infinite or null because there isn't any voltage across it in resonance conditions.

Shielding

Shielding can be used to minimize interference. It is possible to reduce interference by using a Faraday cage (a enclosure made of metal) or mesh, as well as shield boxes.

Inductors and Capacitors

Inductors and capacitors are two kinds of electronic components which can be utilized for storing energy. Capacitors store energy in an electric field, whereas an inductor stores it within an electromagnetic field.

If you connect two capacitors or inductors together, you've made an resonant circuit. The capacities and capacitances in these circuits determine the extent to which they be able to vibrate at a certain rate when they are connected to the power source (a batteries).

Fuses

Fuses protect circuits from excessive current or overloaded power supply and short circuits. They're also employed in fuses to safeguard the circuit from high current that an device or appliance to manage.

Fuses are tiny , round devices made up of two metal contacts at one end, and a hole on the opposite side through which you can view the inner workings. They are available in various dimensions according to their amperage ratings which refers to the amount of electrical energy they're designed for to flow before disintegrating.

A fuse that is rated at 1A will blow if an electrical current of more than one amp is flowing through it. For instance, if you have three fuses linked to each other and listed at 10A, they'll blow all at once if 20A is flowing through them at the same time!

Switches

Switches can be utilized to turn electrical circuits off and on. Switches are employed in a variety of applications, from basic light switches to high-voltage power switch. They are able to cut off the circuit, or even complete it and are utilized to regulate the flow of current in the circuit by breaking an electrical connection. They can be controlled either manually or automatically.

Practical Circuits

Antenna Measurements

An antenna can be described as a piece of equipment which converts electric energy to radio signals. To assess the performance the antenna provides, you have to determine how much energy it is converting to radio signals.

Measurements of Transmission Lines

Transmission lines are a way to carry the electromagnetic force (EME) across two places along a certain distance. It is composed of two conductors, separated by a dielectric (insulating material).

In measuring the efficiency of transmission lines there are three primary factors you can evaluate attenuation in relation to. frequency standing ratio of waves in relation to. frequency reflection coefficient and reflection coefficient. frequency

Transceiver

Transceivers are a type of two-way radio that integrates receiver, transmitter and control functions in an integrated device (transmitter and receiver reside in one box). While most transmitters employ a crystal oscillator to create their signal, they should also be equipped with an appropriate protection system to protect against high-frequency out-of-band signals or the high voltage levels that are that are generated when transmitting.

Receivers are usually tuned to pick up signals at one or more frequencies by using various crystals. They can also incorporate an RF amplifier that boosts weak signals prior to their ability to be demodulated using the detectors of the receiver circuit.

Duplexers allow both transmit as well as reception antennas to be directly connected to an antenna feedline , without interference between the two.

SWR Meter

As was mentioned in an earlier chapter that an SWR Meter is an instrument that is used to measure the standing wave voltage (VSWR) in the transmission line. A VSWR Meter is a specific kind of SWR device that is specifically designed to be used by amateurs.

The two most commonly used kinds of the amateur VSWR meters include bridge and nonbridge-type. The bridge-type model has two diodes as well as an in series capacitor with the feed line of the antenna.

Ohmmeter

An ohmmeter can be described as a device that is used to measure resistance. The basic idea behind it is that when a flow of electricity traverses an object it produces heat, which changes according to the force of the current as well as its route. The device has measuring elements which convert the thermal energy into electricity. This electrical energy can be assessed by attaching them to an eye or needle in their dials.

In simple terms, an ohmmeter consists of three terminals with the two terminals (A as well as B) and a middle spot (C).

One end of the resistor R1 connects to the terminal A, while the another end goes to the terminal. Similar to that, one end of resistor R2 connects to terminal C, while the other end connects towards terminal A. Additionally, capacitor C may be connected between terminals A and B, or C by itself in case you require only capacitance measurement capabilities without the capability of resistance check.

Voltmeters

A voltmeter can be used to measure the voltage.

A voltmeter is used to gauge the amount of current flowing through a circuit. It can also measure it can measure the voltage of any given point within the circuit.

A voltmeter is a device to gauge the specific cell of batteries. It is also able to measure the voltage across the load the terminals on an actuator, by connecting their terminals direct to the two sides of the load terminal or by adding resistance between the terminals.

Ammeters

Ammeters are instruments used to gauge the amount of current flowing through a circuit. They consist of an electric wire coil that is moved as the current flows through it as an electromagnet. The more rapidly the current moves into the coil the farther from its central location it'll move.

Soldering Appliances

Soldering is an essential capability for all electronics enthusiasts So it's crucial to be equipped with the right tools. Here are a few of the most commonly used soldering equipment and the way they can assist you create secure, safe connections between the components of your circuits:

* Solder Iron: A tiny electric or butane-powered tool that is that melts solder to make joints between two parts.

* Solder: A metal which melts upon heating by melting iron, or any other heat source.

* Solder sucker (or braid) is a cleaning device designed to remove solder off electrical connectors, without damage beyond repair; often referred to as a liquid solder remover.

* Flux: A corrosive chemical compound that is applied to metal surfaces prior to soldering to stop burning during heat exposure. also referred to as rosin core, liquid or surfacing paste.

Resistors and Capacitors to power the Radio Circuits

The use of capacitors and resistors is in various types of circuits. Resistors regulate the flow of current within a circuit, and capacitors store energy within the form of an electrical field. When it comes to radio circuits both resistors and capacitors are utilized for a variety of reasons:

* To alter the oscillator's frequency until it is at the desired frequency
* To stop the flow of current between two points.
* For providing feedback to stabilize amplifiers.

The Sensitivity of Radio Receivers

A radio receiver's sensitivity refers to the capacity of a receiver to sense weak signals. Sensitivity is measured in decibels or dBuV, and is defined as the ratio of intensity of that unwanted signal and desired signal, expressed as decibels (dB). The higher the figure is, the higher your transmitter's power output over a certain range should be. It is also known as SNR (signal-to-noise ratio) can also be used here, however it is not as commonly used as the ratio S/N.

The most commonly used notation to measure the sensitivity of a product is ST = 10 * 10 (1 + P/N).

This means that if the signal strength is increased to one decibel (1 dB), ST will increase by 10 dB. If you increase the power of your transmitter output ST increases by 20 dB, etc.

Band Width and Selectivity of an AR Receiver.

Selectivity refers to the capacity of a receiver unwanted signals. The smaller the band width is, the more selective the receiver will be. Band width is the frequency range that the circuit or antenna could take in.

Mixer Circuits and Oscillator Circuits

A mixer circuit utilizes several frequencies in order to create the third frequency. It is employed in radio transmitters and receivers to convert the frequency of the signal.

Mixer Circuit

A mixer is comprised from three or more amplifiers (mixers) set up in a modulo-N arrangement. Each amplifier is equipped with its own local oscillator that transmits an RF signal with the fixed frequency, the f L.

Input signals like V IN1 V IN1, V IN2, as well as V IN3 are simultaneously applied to each input terminal for all mixers. But just one of the opposing inputs is active at any given time. This decides which output signal will be generated at each mixer stage of the circuit.

In general terms If V IN1 is zero volts, then V OUT1 is F LO - f and if V IN2 equals zero volts then just V OUT2 = fLO + f IF both V IN1 and V IN2 equal 0 volts, only V OUT3 is 2f LO + f IF .

Signals and Emissions within the Amateur Radio Station

There are numerous types of signals that are used for amateur radio. Certain signals are able to be heard using a basic device (such as an FM or AM receiver).

Other signals, for instance satellite signals require more specific equipment. Radio stations are an expression used to describe the equipment used to transmits and receives radio signals. This can include antennas, amplifiers power supplies, lots more!

Telemetry as well as Telecommand within Amateur Radio Station

Telemetry refers to the transmission of information about measurement and control from the ground to spacecraft, aircraft, or any other vehicle, particularly to control equipment remotely aboard.

Telecommand refers to a single-way signal to activate, modify or end the function of an electronic device from an extended distance.

Telemetry is a crucial connection between the air and ground. This is why it is employed in numerous sectors of aviation as well as space flight.

Satellite Tracking Beacons and Programs

These programs can help you determine the satellite's location and monitor them as they change their orbits around earth.

Beacons

These are signals that are transmitted through a radio station, which allows you to determine the strength of your signal and decide whether the signal has reached the person you want to send it to.

Modifications for Downlink and Uplink Modes

The uplink and downlink methods of satellite communication are two distinct methods. If you send a signal through the earth to satellites, the method is referred to as uplink.

If a satellite transmits signals towards the earth it's known as downlink. Radio stations that are amateur typically utilize downlink and uplink to communicate.

Spin Fading and "LEO."

Spinfading is a phenomenon that happens when the Earth's magnetic field gets disrupted through solar radiation. The resultant interference causes the solar signal of the satellite to fade and out.

LEO is a shorthand for Low Earth Orbit, which signifies that these satellites are in an elevation of around 500 km (300 miles).

Radio Direction Finding

Direction finding in radio, also called "RDF," is a type of direction finding (DF) that is used to determine the location of a transmitter radio source, like radio transmitters, by taking note of the direction from where radio waves received arrive.

RDF is compatible with any kind of signal transmitted (from AM radio stations, longwave time signals, to radar systems and FM subcarriers) by using radio receivers.

This type of location-based service may also be utilized in reverse by police agencies as well as military units to monitor the movements of vehicles or individuals by using their mobile phones. Modern techniques may employ several antennas that are spaced apart, so there are multiple phases centers accessible to receive signals from different places within their vicinity.

Linking Radio Stations via the Internet

You can connect radio stations via the internet. This lets you connect to other radio amateurs at the same time and without traditional means for communication like telephone, fax machine or even mail.

Exchanging Grid Locators

Grid Locators can be used to define a place in the world. They are composed of six numbers, each representing the distance to one of the 60 points which make up the grid. The grids are placed 10 minutes apart and located around the Equator. The first number indicates the distance north or south you are from one of these locations, also known as meridians of reference (for instance that if you're located at 40degN, then 0 stands the number for Greenwich).

The second number gives the east-west location measured in degree (0 being used for Greenwich) The the numbers 3 to 6 indicate the longitude east or west beginning at zero degrees (European) or the 0degW (American).

Image Signals and NTSC in Amateur Radio Station

NTSC is the acronym as the National Television Standards Committee. It was

NTSC was the color television standard that was used throughout the United States and other countries up to the mid-1990s.

PAL is the abbreviation to mean Phase Alternating Line, a standard for video broadcasting that is in use across many countries and regions around the world.

SECAM (Sequentiel Couleur Avec Memoire) is another digital television broadcasting standard that was developed in France and employs an entirely different method of coding colors than NTSC as well as PAL systems.

RGB is an add-on color scheme that makes use of the green, red as well as blue components of light to create the other colors that are that are visible to human eyes. It's just one of the options for how colors can be generated using a computer's screen or TV display device.

Morse code (CW)

CW is the most earliest mode of radio amateur, and it makes use of on-off keying in order in order to encode text messages. It's been utilized since the beginning of radio, however it's still utilized for text messages even today.

Packet Radio

Packet radio employs an electronic format to transmit messages in text over the airwaves via a kind of modem known as An AX.25 network.

It is the AX.25 Packet System was designed to work with radios that use HF (shortwave) radios. It can also be used with UHF/VHF frequencies as well as satellites such as AO-27 SO-50, AO-27 and many more. You can also broadcast with Winlink 2000 or similar software applications from your computer!

PSK31

PSK31 is a digital type of mode which transmits at 31 baud, instead of 19 baud that is typical for the majority of modes such as RTTY as well as ASCII (text). It was created by Friedrich Bauer in Germany in 1992 following his experiments with different methods of transmitting more quickly than standard morse codes.

The program doesn't need any additional equipment other than your computer with a sound card program like Audacity as well as Fldigi.

Error Detection and Correction for Amateur Radio Station

Correction and detection of errors is an essential to radio data transmission. The ability to detect and correct errors refers to the ability to spot mistakes in data and rectify the errors. It is able to detect the presence of noise, frame synchronization and packet sequencing. It also allows estimation of bit error rates.

Correction and detection of errors is essential when applications require data transmission across long distances, or in situations where the characteristics of the channel are not predictable.

Correcting and detecting errors can be utilized in a variety of applications, like the storage of data, memory in computers as well as digital communications systems that employ error-correcting codes such as the cyclic redundancy check (CRC).

RDF, also known as radio direction finding equipment RDF could be utilized to find the person who is sending out a signal. This is done:

* You'll need your compasses and a map of location of the person.

* Make sure you know your exact place on the map so that you know which location they are in relation to you.

There are many ways we can determine the signal coming from them, or not:

* If it's located situated on one side of us and not the other, then it may be within our view however it is not right above nor beneath us (in the latter case, we'd need to know more).

Another method to tell us whether the information is directly from them is to use triangulation which is the process of the measurement of angles between different sources.

FM as well as SSB Modulation Circuits

Frequency modulation (FM) and single side band (SSB) are two different types of amplified modulation. What is the difference between the two and what circuits are utilized to modulate the frequency in these modes?

Frequency Modulation (FM)

FM is a term used to describe frequency modulation, in which the frequency of a high-frequency wave is dependent on the current value of the input signal. Also If an FM signal exhibits positive values at any point during its entire cycle, it's called "modulated" on the particular carrier at the time of.

The greater the amount attained by this modulating signal throughout its cycle the more powerful its impact will have on the amplitude of the carrier and hence, we can say that the FM system is based on frequency division multiplexing (FDX).

An effective way to describe FDX methods is to think of them as consisting from two subcarriers which transmit information at the same time, but in a different way. As long as we keep in mind that each subcarrier could include multiple carriers (and therefore, a variety of channels) within it!

One Side Band (SSB)

The term is specifically referring to an aspect of AM that is typically found in HF communications links, when conditions are bad enough to prevent CWR or FSK operating modes to be used continuously over extended durations.

Amateur Radio Antennas and Feed Lines

Feed Lines

They are wires that connect your antenna to the receiver or transmitter. The feed line may be constructed from any kind of wire, which includes copper, braided, or even twin-lead (the type used in many wireless antennas). Another kind of line for feeding is coaxial cable. Another variation of this concept is the ladder line that is insulated with dielectric between its central conductor wire as well as its shield braid.

If you're considering commercial radio stations you might notice that they utilize "house wire" as feed lines!

As we have seen in the previous chapters, there exist many different kinds of antennas. But there are some antennas that do not require specific feed line, as the antennas are made to feed directly into the amplifier, or into an impedance-matching network (IMN).

Attenuation vs. Frequency

Attenuation is the term used to describe the amount of signal lost in the cable. It is expressed by decibels (dB) or as the ratio of power input in relation to power output. Attenuation is a crucial factor when choosing the feeder lines to your system, as the smaller it is the greater distance you can transmit without having to include an amplifier in order to increase the signal's strength prior to sending it out on the airwaves.

Attenuation comes in two components two components: dielectric loss and resistive loss.

Resistive loss happens as electrons pass through conductors. The more conductors are there in a cable, and the longer they run in the cable, the more intense this type of attenuation is likely to become.

Dielectric losses are caused by the insulating materials such as plastic. As an example, imagine that there's a lot of insulation in one location with respect to another area in your circuit board. In this case it can cause greater that normal losses in dielectric, which could result in an inefficient performance and decrease the life of your circuit board.

Antenna Tuners

To receive and transmit radio signals The impedance of your antenna has to be compatible with the radio's impedance receiver or transmitter.

The term "matching" implies that the two devices are able to transfer energy effectively, and neither device using up energy because of transmission losses.

The impedance of an antenna could be described as its resistivity (R) in units of length (L). It is measured in units of ohms (O) the same way as resistors are measured in ohms (O). The formula used to calculate O is:

$Z = R Z / (p 2 L)$

In this equation, Z is the impedance, R is the resistance in ohms. the equation p is 3.141592654 ..., f is the frequency in Hertz (Hz) while L represents the distance in meters.

Horizontal and Vertical Polarization

To better understand the distinctions between horizontal and vertical Polarization, let's begin by examining the characteristics of their propagation.

Vertical polarization is the preferred method of transmission in the band HF (3 between 30 and MHz) which encompasses both data and voice transmissions, as it reduces interference from sources of noise. Additionally shortwave broadcast stations employ vertical polarization to transmit signals across long distances.

Vertical antennas can also be used to communicate with satellites via line-of-sight because the distance of satellites from earth hinders horizontal antennas to detect the signal.

It is worth noting that when you use handheld radios at lower frequencies like two meters, or even 70 centimeters (VHF) it is possible to expect to get more success with an antenna that is polarized horizontally rather than one that's vertically oriented.

Antenna Gain

Antenna Gain is the ratio of energy radiated by an antenna to the power that is supplied by the antenna. It could also be a measurement of how effectively the radio receiver or transmitter broadcasts radio signals.

Antenna Gain measures how effectively an antenna radiates out signals into space. It is expressed as decibels (dB). The calculation is done using the following equation:

[power radiated]/[power supplied] = [antenna gain]

A successful Ham radio setup requires not just a reliable receiver and transmitter, but also a reliable antenna system. The most crucial component of an antenna setup is ground. Without a strong ground, you won't be able get the most effective outcomes with your antenna.

It must be connected to a solid earth ground attached to the power supply of your station and not to any other item that is grounded using another wire through the home.

The choice of an antenna

If you are choosing an antenna to usage on the 10-meter range, it needs to be around 48 feet long and resonance with 28 MHz. This antenna will work to use for CW (Morse code) However, it is not suitable with SSB (voice).

An alternative in this instance is an 80-meter dipole, which could be made out of two 40-meter dipoles joined in their central insulators with sections of wire cut out of RG58 coaxial cable, or other thin wire strands.

Hf Yagi Antennas

Yagi antennas are directional ones which use one active element in front of an array of components. They are widely used in HF and shortwave radio. They are utilized for communication over long distances. They are used to transmit messages across oceans, through mountains and even through tall buildings.

The Yagi antenna is made up of three pieces:

* A actively active component (the one that is closest in proximity to your),

* A variety made up of parasitic components (the elements that are on either side) and

* Reflector elements (the one with the farthest distance).

Each piece's length differs in accordance with its frequency, however they all are based on the same principle. There's a central part of the lobe that you would like your signal to be focused on, and it will send the direction you want it to go while cancelling out any noise coming from other directions with methods of phase cancellation.

Mobile Verticals of HF

Install a vertical antenna on your mobile station then you're able get signals in all directions. This is ideal in DX or long-distance communication.

Vertical antennas also have more gain than a dipole This means that it can transmit further than the same wire, which acts as an antenna by itself. Therefore, vertical antennas are typically utilized alongside ground planes. This increases both performance and effectiveness.

The ground plane is a part of the "unbalanced transmission line" which is necessary to counteract loss within the feed line as well as coaxial cable which connects the antenna and the radio. It also helps protect against lightning attacks by creating an electrical shield over your vehicle (or your home).

Portable VHF/UHF Yagis

The Yagi antennas are directional ones that are suitable for mobile and long-distance communication. They can be used for VHF (30-300 MHz) and UHF (300 MHz to 3.0 GHz). The yagi antenna is among the most well-known of all amateur radio antennas because of its superior increase over a dipole or vertical as well as its affordable cost.

The yagi is composed of some or all of the elements which are linear in their arrangement and have one end positioned toward an arrival direction, and the opposite end facing it, creating an array that is able to maximize signal transmission in a certain direction.

Fixed Beam Antennas

Fixed beam antennas were made to emit in a single direction. Fixed beam antennas can be utilized for fixed stations, long-distance and satellite communications. They are also utilized in amateur radio, military radio and commercial radio communications.

The Dipole Pattern

A dipole pattern can be described as a graph of the pattern of radiation that is characteristic that is produced by a dipole half wave. Dipole patterns are a crucial concept that amateur radio operators must comprehend because it describes the way that an antenna emits electromagnetic energy. It can also be utilized to determine which kind or element (such as coaxial cable or wire) is required for feed lines that carry the signals from transmitters to the intended equipment for reception.

Dipoles are commonly used as resonant antennas on LF or MF frequencies since they are able to be constructed with a minimal physical profile when compared to other kinds. This makes them perfect to use on ships and aircrafts where space restrictions may be a problem. They are also less likely to be able to detect interference as effectively than other types and, therefore, if you wish to allow your station's signal reach further out into space, without having issues with noise it could be useful!

Antenna Loading On Ham Bands

Antenna load is the proportion to an antenna's impedance its characteristic impedance the feed line.

An antenna's impedance is determined by its structure and the location of it and location, while the feed line's impedance is determined by its physical characteristics and length. The relationship between these two impedances determines the much power is transmitted from the radio to the antenna at any particular frequency.

A property that is frequency-dependent determines the impedance characteristic of feed lines (usually 75 Ohms) known as the velocity factor (VF). The VF value can be calculated in the terms of delay in propagation in the following manner:

VF = = 2 * 2*3 * 5 * 2 *33 30 (1 + 1) 1/[2*3])/[(1 + 1/(2*3)1)) + 4[4

Where:

* is the wavelength in meters at the operating frequency.

* * 2 * 5= 10 Upper Half-Space Dimensions per meter; Two lower half-space measurements equal to this number

therefore you are able to not pay attention to them right now (they do not cancel each other out)

Large antennas, high power and reliable feed lines are essential for a high-quality quality reception as well as transmission. Many factors influence the effectiveness of an antenna as well as the feed line, which include:

Antenna Power Gain

A antenna's strength is the amount of signal it can receive than is received by an isotropic radiator (dish) with the same dimensions.

Antenna Radiation Efficiency

The efficiency of radiation can be defined as the percentage of the radiation power from the antenna to the input power that is supplied to the antenna.

Security of Amateur Radio Station

Similar to any other pastime that involves electric power, the hobby of amateur radio is also prone to some risks. In contrast to other hobbies that require electricity

However amateur radio is able to provide the potential to be an avenue of communication with other agencies in the event of an emergencies.

In this way, Amateur Radio is regulated by various governmental agencies across the globe. Different regions have distinct rules and regulations, but all of the bodies that regulate them all require that users take a safety test before they can operate an amateur radio station.

Circuits and Hazards of Power Circuits and Hazards

Circuits for power as well as circuits for RF (Radiofrequency) circuits must be separated within an amateur radio station. The reason for this is that If the two are mixed , they could create interference or harm your equipment. That means power cables shouldn't be run near coaxial cables carrying radio signals, and should not be used close to any device that is that is connected with the radio transmitter.

It is also important to maintain your power lines as often as you can by making sure that nobody gets in contact with them while they're utilized (i.e. using their hands). This will increase security for all those working in the facility and decrease dangers to the equipment Therefore, you should keep people informed whenever you can!

Your antenna could create problems if it is not secured or grounded correctly. It is therefore essential that all the components accessible at ground level are grounded prior to putting them into your antenna setup (for instance computers, for instance).

It is possible to have something that isn't conductive such as plastic under the area where someone is standing during the operation. Make sure it doesn't break when time passes, since once it's damaged it will not provide adequate protection any longer.

RF Hazards

RF can cause burning sensations on the skin when it is exposed to a powerful or high-power field of RF. This usually occurs due to being exposed to UV radiation which are invisible lights.

Exposure to radiation

Radiation exposure is among the main safety concerns of amateur radio. The radio waves are absorption from the body and then reflected by it, then scattered by it. The body may also reflect it.

Proximity to Antennas

The close proximity of an antenna towards the body is identified as causing burns from radiofrequency. This is preventable with a careful placement of your antenna far from your body, particularly when you use high-power.

Risky Voltages

When working with live circuits, it is important to ensure that you treat these circuits with care. Always ensure that the circuit that you're working on is powered by voltage, and take the necessary precautions to avoid personal injury.

Be careful not to handle any terminals or wires unless you're sure there is no voltage. If you are uncertain about the state of a circuit, you can use the voltmeter to determine voltage prior to touching any item.

* When working with high voltages, it's essential to not rely on your body to be ground conductors for connections or testing equipment that has unknown working voltages (i.e. that you should not apply).

If a person is able to touch both hot and neutral leads at the same time (e.g. walking in bare feet on concrete that is wet). In that scenario the person could be electrocuted even though there is no any direct connection between these two electrical sources! This kind of incident usually happens due to people's inability to consider the proper way to ground their bodies when working near high-voltage lines. If you're not aware of how your body behaves in the role of an electric conductor when working with these kinds of electrical lines it could lead to serious injuries or even death!

Lightning Protection

Protection against lightning is mandatory for every amateur radio station. This applies to an antenna and coax connectors feeding line, operating position and the equipment within the station.

Lightning protection is available as a complete set or constructed using materials easily accessible at home centers and hardware stores. But installing lightning protection according to the industry standard along with local code of building regulations is essential. Typically, this involves that it is placed on the roof's top that is not accessible by other people (i.e. it is the roof isn't an overhang) or on an exterior wall close to where other structures are located (i.e. chimney) or even in an interior wall close to the point where electrical wiring is accessed through external sources, such as electricity lines (i.e. basement).

Grounding

It is a measure of safety. It also helps to eliminate static electricity, lightning energy the energy of residue, as well as extra voltage. Additionally, it helps stop any dangerous arcing and corona releases that might be spotted in your device.

Grounding can also help prevent injuries from shocks for workers who work on equipment that has grounded chassis. This could happen if one person is in contact with the live conductor, and another person comes in contact with the same piece of equipment prior to it is discharged through the isolation transformer, or any other device that releases all charges that accumulate all at once (a earthing resistor). This is the reason it's crucial to properly ground radio stations for amateur radio so that they don't cause any harm or injury!

Circuit Breakers and Fuses

Circuit breakers and fuses are made to shield your from harmful voltages. They differ in that they guard against various dangers, but they perform the same task to block the flow of current when they get too hot or are not conductive.

Fuses have an earlier melting temperature than circuit breakers, which is why fuse are often used in applications with high current like heaters or electric motors. Circuit breakers too have lower melting points, but can be reset in the event of no permanent damage from breaking off or melting down.

With careful planning, the risk of accidents are reduced to zero. Thus, safety is everybody's obligation. It's tempting to think security is the responsibility of the chief communications officer or engineer. However, in actual it's the responsibility of everyone. So, when looking at safety, we need to keep in mind that it's a continuous process, not a final destination.

Chapter 12: Commission's Rules In Amateur Radio Station

A radio amateur is someone who is named in an amateur operator or primary license granted on the FCC ULS database. It is the ULS database serves as the primary FCC database that contains information on licenses for all radio-related services. If you find your name listed there, and it is tagged by a call number that indicates you are an amateur radio operator.

A station that is part of an amateur radio service that is comprised of the equipment required to transmit radio signals are called an amateur radio station. It could be a handheld radio or a base radio or a base radio.

The reason why the amateur radio service is there?

* There's a lot of real estate available in the radio bands. Some could be sold at a price of million of dollars. But the FCC grants the spectrum to us as it is aware of and promotes the worth of the amateur radio

service for the public as a non-profit, voluntary communications service. EMCOMM (emergency communication) is a key element of this.

* Continued and extended use of the ability of amateur radio to aid in the development of radio science. The radio technologies you use today originated in amateur radio , because we have the ability to do lots of experiments and not be in conflict in the FCC.

• Encouragement and enhancement of the amateur service by means of rules that improve abilities in both technical and communication aspects of the craft.

* Expansion of the current reservoir of the amateur radio service that includes qualified technicians, operators as well as electronic specialists.

* Expansion and continuation of the ability of amateur radio to increase international goodwill. We are able to represent ourselves on the world stage through radio amateur.

The FCC is the body that makes rules and enforces them for amateur radio throughout the US and its possessions. Additionally the FCC issues licenses to amateur radio operators.

Two teams administer exams for amateur radio licenses . One team is the volunteers who coordinate the examinations (VEC) which is an organisation which the FCC is appoints to conduct examinations, papers as well as everything else required. VECs serve as a bridge with both the VE (Volunteer Examiners) and the FCC.

Volunteer examiners (VE)

Volunteer examiners are amateur who is accredited through any or all VECs that offer to conduct the amateur license test. If you take your test, you will require at least three examiners present during the test. They must all have an amateur license that is general or higher to take tests for technicians. Also, if you take an amateur radio exam There three examiners who will observe you, give grades, and make sure that your paperwork is completed.

The station owner is responsible for the operation of your station in accordance with rules of the FCC rules. The license for amateur radio granted through FCC serves two purposes. FCC serves two functions:

It's your operator's license. It's affirming that you've done all the things you're required to do and that you're licensed in the correct manner.

It's your primary station

A single person can have one primary station operator license. You cannot hold more than three amateur radio licenses. It is one license only.

Check that your mailing address is up-to-date on your FCC ULS Database. If the mail you send is returned to the FCC as indeliverable your license could be suspended or revoked. This is due to the fact that the FCC must be able locate your. Furthermore, once you have obtained an amateur radio license, the FCC has the right to examine the station of your choice at any point. Also, they may examine your equipment, and only it at any given time.

The term for a license to operate amateur radio is 10 years.

It is possible to renew your license for up to 90 days before the time when your license is due to expires. In the event that you do not renew your license for any reason, you'll be granted an additional two years of grace however you can't be operating during the grace period. However the FCC keeps all your data which means that all you need to do is renew. If you do not renew after two years, you'll have to start again.

The places where you can send

After you've obtained the license, you can operate wherever you want. FCC is able to regulate:

* The continental US
* Alaska
* Hawaii
* Guam
* Puerto Rico and
* The Virgin Islands as well as other countries that have bilateral operating arrangements.

You can carry your radio to any place within Europe and use in accordance with their regulations.

No matter what the language that you use, you need to identify as English. Also, you should use whatever language you like but be sure to identify yourself in English.

After you've been issued a permit and a license, you are able to be a part of international waters or airspace aboard a US-registered vessel or aircraft. But, you need permission from the captain of the vessel or pilot of the aircraft who is in charge.

Amateur radio call sign

The call signs are distinctive - there is no exact match and the style of the call sign is based on the type of license. The call sign is a way to identify both the station as well as the operator.

A 2 3 call sign two letters as a prefix as well as a number. It also has an suffix of three letters. The prefix can consist of two or one letters in all cases. The first letter of this prefix within the United States will always be an N, K, or. District numbers ranges from zero to nine based the area you live in.

In the picture above in the image above, the correct answer is W3ABC. The first answer doesn't match the format, while the second one doesn't contain an digit and the third one is the only number that has an a.

The FCC issues licenses through sequential orders. Technicians who are first licensed first receives a sequentially issued 2x3 (2X3) type station-specific callsign. You aren't able to select the call sign. Your location determines the area that will issue the call sign.

One person can only have one primary license or station license for amateur radio. An amateur radio operator could serve as a trustee for several club licenses. A club license must be held by at the very least four members.

Special event callsigns

Any licensed amateur operator with FCC license can request a temporary one-by-1

(1X1) specific event sign. Special call signs for special events are issued for a limited time - typically for the duration of the event. After that, they are reused. You can also include self-assigned indicators on your sign to provide clarification or to provide more details.

If you're upgrading your license your license, you should include an indicator on your call number while waiting for the new class of license to be listed on the FCC ULS If you are using the new frequency privileges:

Callsigns for tactical use

In these instances, communication is more effective when you have a functioning identification number of your place.

It's not unusual for a radio amateur club or other organization to offer communications for marathons and, for instance, storm spotting or something similar. So instead of keeping a list of names and figuring out the race headquarters, rest stop 2 or EOC you could simply dial race headquarters or rest stop. EOC or weather service or whatever the name is. This way, anyone who is there will listen to "race headquarters" and then answer the radio.

If you're working under the tactical call sign, each time 10 minutes, you'll need to verify your identity using the actual call sign.

Switching your call number

When you receive your first call sign generated by computers, you could be eligible to make it one that is that is more meaningful to you by using this vanity sign application. Similar to the way you can get customized plates for your car and other vehicles, you can also purchase an individual call sign. These include callsigns with your initials, or call signs that represent particular interests or deceased relatives or a loved one, etc. But, the condition is that the sign must be in place and you cannot take an existing call sign from someone and it must be accessible. You can request a one-by-3 (1X3) or even a two-by-3 (2X3) decorative call sign for tech.

A 1X2 and 2 X 2 can be used additional class licenses.

Why is it that the FCC issues callsigns?

To identify the station or operator who transmits. Therefore, you need to be able to identify by sending your call numbers each time you transmit. In accordance with the FCC regulations, you have to identify using two methods:

* At the conclusion of a conversation and
* Once every 10 minutes

In this way, you may engage in a long conversation for an hour and at least once at every 10 minutes you can identify by sending your call number. This is also the case when testing your antenna or equipment.

Extra requirements applicable to events at special venues:

Note the location from above

* The operator of the radio must be identified with their call number at least once every hour.

Although the FCC regulations don't oblige it, it's an ideal practice and courtesy to be able to identify yourself each time you contact other stations.

If you would like to join an in-going conversation (known as QSO) between two or more stations, you must wait for a timer between transmissions and make your call number one time. Then, the other stations will usually acknowledge your presence and permit you to join in on the conversation. However when they're having an intense discussion they may take some time to acknowledge you, but generally Hams are adroit about inviting you to join.

Phonetics

You must be aware of the phonetic alphabet as well as its usage in voice communication. It is the International Phonetic Alphabet (IPA) is a spelling system that uses words using 52 fundamental symbols to represent the sounds of speech. It is commonly used to spell out names of places, people, as well as objects used in the voice communication. In the Amateur Radio station phonetics can also be spelled using phonetics using the IPA alphabet.

Radio operators who are amateur utilize their International Phonetic Alphabet (IPA) to transmit messages using Morse code or radioteletype messages during radio contact.

There are more than 100 countries that have recognized this alphabet as a standard to represent speech words or writing down the words on paper so they can be easily read without confusion as to how they were initially pronounced by those who speak these languages.

The phonetic alphabet that is used worldwide to make phone (phone) messages is displayed in the image below.

Each alphabet has its own identification number, and none is identical to another. This was made a rule by the international telecommunications union (ITU) which is accepted as an alternative to alphabets.

The FCC recommends using phonetics, particularly if you're a novice Ham. Be careful not to use silly, random terms or phrases to identify your station, as people may not understand them. So, stay as close to regular phonetic system.

The ITU

The ITU is a global organisation that creates guidelines and frequency rules. Every country on earth is part of one of ITU regions. Three regions are divided across the globe that assist in managing frequency allocations. It is located in ITU Region 2. United States is in ITU zone 2.

Your duties

Your obligation as the station's licensee is to ensure that you operate according to FCC rules. If you transmit with amateur radio equipment You are the control operator. You are the one in charge of the equipment.

Every radio station that is amateur broadcasting has to have a controller and someone has to be responsible for the equipment.

The operator licensed to manage the transmissions of the station is known as"the controller.

A control point can be described as where at which the operator's job is carried out.

Sometimes, the point of control could be remote. For instance, a number of the latest radios have the capability to control the head the transmitter and its body via a network, referred to as remote control.

Also, in some cases the control point could be distant from the transmitter, and station control is carried out via a telephone line or radio link.

In some cases, the radio transmitter operates independently, however, an operator is able to take control at any moment via a telephone line or a separate radio link.

The operator of a control station at an amateur station restricted to the permissions that they have to operate under their license. You can't gain privileges by visiting the home of a different ham or to a control station. Thus the technician class licensee running the equipment of an additional class licensee is still entitled to the privileges of a technician class and vice versa do not gain rights for making use of equipment belonging to another ham. In this case the operators of both stations are accountable for their operation in a safe manner.

Reasons for and Use Permitted by Amateur Radio Service and its Purposes Amateur Radio Service

The Amateur Radio is an passion and an activity. Amateur radio is a non-commercial, voluntary communications service that aims to promote international goodwill, technological advancement and public service initiatives.

Amateur stations broadcast at frequencies specifically designated for use by amateurs by the International Telecommunication Union (ITU) Radio Regulations.

Operator/Primary Station License The grant The grant of a primary station operator license is the granting of an operator's license to someone who is who is responsible for the operation of an amateur station.

In some cases the primary station license grant also confirms the holder's access to an auxiliary operations, or repeater operations.

Definitions of Terms of the Basic Types Used in FCC Rules

Following terms and phrases are defined by SS 97.3(b) (b) of FCC rules:

Amateur station. Radiocommunications are conducted with amateur radio radios to train self-teaching, intercommunication and technical research conducted by radio amateurs.

These stations could be considered experimental, but they're still amateur stations if they operate at frequencies higher than 50 MHz.

Amateur operator. The person who is named in an amateur operator or primary license granted on the ULS licensed database that is consolidated.

Operators need to have either the Technician Class licence or a higher class operator license, and be able to pass an exam administered by the Volunteer Examiner Coordinator (VEC).

ARES (Amateur Radio Emergency Service)
ARES (Amateur Radio Emergency Service): This group provides the ability to communicate directly between various local entities like police departments, hospitals and fire stations in the event that normal communication systems fail, for instance as during extreme storms, such as hurricanes, or when power outages result from blackouts due to earthquakes.

Interference
Interference refers to any electromagnetic energy in radio frequencies that disrupts or blocks, or in any other way reduces or hinders the communication system's.
To prevent interference Amateur radio operators need to comply with the following guidelines:
• Know the type of equipment they're using and how they operate it.
* Only use legally licensed frequencies within their region.

Use antennas that don't interfere with other stations or other users at adjacent frequencies by using interference control for radiation (antennas that have a higher gain could create substantial interference).

It is best to avoid placing an antenna over structures in urban areas from the fall and late spring, because this can increase radiation to other buildings close to your site. Also, it makes it difficult to detect weak signals because of the electrical noise that is generated from raindrops striking the antenna's surface in storms.

Rules for RACES Rules

RACES is a nonprofit association made up members of radio amateurs that have been trained to assist in emergency communications in emergencies or during disasters. RACES stands for Radio Amateur Civil Emergency Service.

The RACES license is not a distinct type of license. Instead it refers to people who have current FCC-issued amateur licenses , and have completed the mandatory classes that are required to be completed by the

ARRL Section in their local area. ARRL Section to participate in the program.

Communications - the Cornerstone of Ham Radio

Amateur radio is built on communication with other radio stations within your local region and across the globe. The communication is conducted via data, voice photographs (including television) and various other ways.

There are several ways to connect using amateur radio.

* Voice - Use your voice to converse with other amateurs via the air or through repeaters

* Data - using computers that are connected to the internet to transmit messages back and back (email)

Image - sending photos or videos over the air

* Some amateurs are using digital modes, such as PSK31. This enables them to talk even when they are far away from other users who have mastered digital modes.

Frequency Allocations

A treaty between the United States and other countries allocates Amateur Radio Service frequencies. In turn three frequencies are reserved to be used by amateur radio operators including the VHF band and the UHF band along with the microbands.

The VHF band includes radio frequencies that range between 30 MHz to 300 MHz (300 KHz and 3 GHz).

The UHF band is comprised of radio frequencies that range between 300 MHz to 3000 MHz inclusive (3 3 THz to GHz).

Emissions methods

To transmit information radio stations use emitting modes. The most popular methods include:

* FM
* AM
* SSB (Single Side Band)
* CW (Continuous Wave)
* RTTY (Radio Teletype)

Information encoded

It's clear that encode information is an essential part in Amateur Radio. It's actually the only thing that's important!

The most commonly used modes by amateurs include AM (amplitude modulation), FM (frequency modulation), SSB (single sideband) as well as the CW (continuous waves).

Each type of mode has advantages and drawbacks. The best method to determine the best option for your particular situation is learning what each mode does well and the areas where it may have issues.

Methods of communication

The modes of communication refer to the various ways that information is encoded and transferred among two radio stations.

There are a variety of modes of communication, however the most commonly used are digital, voice CW and packet.

Spectrum Sharing

Radio operators who are amateur radio operators should be aware that they share radio spectrum, and behave according to this. Thus, interference with other users is forbidden. The next chapter offers guidelines to prevent interference as well as how to handle interference from other users.

FCC Rules

The FCC issues licenses for amateur radio and the rules we follow as amateur operators are contained in the part 97 of FCC guidelines and rules. You can visit the search engine, search for 'FCC part97 to download the entire PDF.

The FCC is the body that sets and enforces rules and regulations that govern the amateur radio services within the US. They are able to check your amateur radio station at any point. But, they will not just visit your home in the event that you're making a nuisance.

The radio amateur licensee is in all times responsible for the functioning of their radio station as well as equipment as well as guarding against misuse.

When you're at home, you may disconnect the microphone as well as the cord for power,.

* When you are in your car Remove the microphone in your car.

If it's portable, make sure to keep it locked in a cabinet or drawer, or in a secured area where possible sources of interference are not able to be able to access them.

Things that are not permitted in amateur radio

Sending music: Consider your surroundings. For instance, incidental manned spacecraft music is fine.

broadcasting in general to all people The public is not all of the public. Anyone who has an audio or scanner can listen to on the radio to listen the broadcast It's not private.

The transmission of ciphers or codes (encryption and secret communications) is not permitted. It is however acceptable for satellite control or spacecraft only. For instance the time that amateur radio satellites transmit the data, they are permitted to encode their data since they don't want any one to be in control of the spacecraft.

Intentionally causing interference It is unlawful to block a repeater on your radio, preventing other users from using the repeater. Intentional interference can lead to an inspection by the FCC more quickly than any other.

Unidentified transmissions: Controlling RC models is fine but "kechunking" the repeater isn't. If you choose to key-up the repeater, you have to provide your call number - even if you're just doing an inspection of the radio.

Use of vulgar or indecent language: Don't use vulgar slurs about ethnicity or race or off-color jokes. If you would never use it in front of your grandma or mother and don't use something on the air. There isn't a list of official definitions of words that are considered to be harmful.

Sending deceptive or false communications such as gossiping or talking about someone else to cause harm.

Communications for business: No communications in connection with your work. But, occasionally, radio equipment from ham radio "swap nets" are permitted. "Swap net" means that you announce the radio equipment that you are offering available for sale, in order for other hams to contact you.

Interest on property There are some instances where it is not:

* Teachers who utilize amateur radio in their classes.

Public health officials who utilize amateur radios to provide hospital-wide communications outside the hospital in emergencies.

* Fire, police, as well as emergency personnel that have amateur radios in their emergency operations centers (EOC)

In essence the hams (hams) have the right to talk to each other. Two-way communication and one-way communication are permitted. One-way communication is used to test devices, RC model boats, robots, aircrafts beacon stations, report on the position. If you're planning on controlling an RC drone or robot, the transmitter box you use will have an inscription with your call sign and you're only allowed one Watt of power. Always identify yourself the call sign with your name.

Two-way communication with other hams on amateur frequencies don't include:

The general population

* CB operators

• Other services for radio

One exception is that every year, there is an Armed Forces Day communications test. Thus, communications with the military during this period are permitted.

Transmitter power

In order to not exceed the maximum power allowed on a specific band, make use of the minimum amount of power needed to complete the communications.

The majority of bands have a power limit of approximately 1500 watts. Of course, certain down in the HF bands are to less however 1500 is usually the highest power rating.

If you are using the minimum amount of power that is required to get the minimum power, you will gain two advantages:

* It decreases interference

* reduces the power consumption of portable radios

Radio communications

The fundamental structure of transmission is illustrated in the above diagram. The microphone first transforms voice signals into electric signals. The transmitter then transforms these electrical signals into radio signals and after that, the antenna emits radio signals.

How do you prevent interference with another station

1. Do not transmit if you are concerned that your transmissions will create a harmful interference for other stations.

2. If you are unsure whether that your transmissions could cause interference, test them prior to sending them out over the air.

3. Be aware that if it's discovered that your communications resulted in negative interference, you could be subject to a monetary fee from FCC.

4. Be aware that if someone is complaining about your actions causing an unintended impact on their receivers and repeater devices (such as blocking their transmissions) the report could lead to an enforcement action against you!

5. To prevent being impacted by an operator, do not use channels that are already occupied such as simplex frequency during local nets.

Transmissions near edges of bands

Transmissions that occur near the edges of bands are illegal, hazardous and are a risky idea. They could cause interference to operating radio stations from other stations. They could also interrupt emergency communications, and interrupt services provided by amateur radio to aircraft during flight.

In certain situations, transmitting outside of your permitted amateur allocations may be a cause for being prosecuted by law enforcement agencies.

Inquiring about for the International Space Station (ISS)

For contact with for contact with the International Space Station (ISS) it is necessary to find the orbital location of the ISS. The best way to do this is by checking an online database such as [https://www.heavens-above.com]

You might be seeking to contact the ISS. In this case you'll need to be aware that it revolves around the Earth every 90 minutes, and is visible for approximately 10 minutes when it makes successive passes over your area.

The most ideal times of the morning are between 8:00 am until 4:30 pm local time. This is when the majority people in North America can see it as well as Europe as well as Africa being able see it later in the mornings of their respective regions also!

If you'd like to reach them via your home but you don't have equipment or aren't yet ready to get licensed as an amateur radio however, there are many ways you can connect with them through social media! Many Facebook groups are exclusively dedicated to communicating with astronauts on missions.

Power output

The power outputs are measured in Watts. The output power from a broadcaster is restricted to a certain extent by FCC, ITU, and CEPT organisations so the Amateur Radio stations are not nuisances or harming other users of the radio spectrum. Any power levels that exceed the maximum permitted by these organizations could result in license being revoked or suspended.

The FCC restricts broadcasters with amateur radio to the maximum of 50 watts of power output in any frequency band, including the HF, LF, MF and VHF/UHF bands.

In addition to in addition to the FCC regulations In addition to the FCC regulations, as well, the International Amateur Radio Union (IARU) has set additional restrictions for each station they represent. Radio rules for amateur radio differ from country-to-country. These fundamental sections will aid you in understanding Amateur radio band.

Amateur radio is among the most loved hobbies around the world, and there are around 3 million licensed users across the world. Radio stations for amateurs are frequently known as Ham radios, or Hams. The hobby of radio amateurs has expanded to include modern methods of sharing information and entertainment among individuals or crowds of individuals. An understanding of these methods is required for new operators who wish to take on this thrilling hobby.

Amateur Radio Authorized and prohibited transmissions

The Amateur Radio Station is authorized to transmit only using the frequencies and modes that are specified within their licensing.

They must clearly identify their station according to these rules, which includes the station's name when it is called or answered.

It is forbidden to use offensive communication as well as engage in other form of communication that may be insensitive to other people.

They are also forbidden from sending any message which has to do with business, or aids in the promotion of the brand or service, unless expressly authorized in a licence that is issued in accordance with these rules to accomplish that.

Inappropriate language and the compensation for operating

Inappropriate language is not permitted on air. You could be penalized for using offensive language in the air.

The FCC has ruled the use of certain terms is not acceptable and therefore cannot be used publically even over Amateur Radio stations.

These terms include those that relate to sexual actions or excretory functions as well as anatomical body parts (e.g. the asshole or Dickhead, etc.).

Interference

Your station should not interfere with other spectrum users or any other communications service which make up a small portion in the spectrum (e.g. the cellular phone service).

If your station interferes the communications of another user, even if it is not intentional this is a violation of Section 333 of Title 47 US Code Part 97, which bans the creation of harmful interference by any means, which includes imitation or simulation (also called "jamming").

Other signals that are transmitted by amateurs can be retransmitted

You can transmit the signals of other amateurs, however you have to get approval from the radio station that broadcasts the signal you want to retransmit and not charge any fee for this.

So, in the exchange of a fair amount of service you may offer a point-to-point connection to two stations, or with an outside facility that is not your responsibility.

Equipment encryption and sales

1. It is not possible to secure your data transmissions.

2. You cannot offer radio-related equipment for sale.

There are some exceptions to the rules above Low power equipment may be sold, as well as non-commercial radio use is permitted.

Transmissions that are not identified and one-way transmissions

"Unidentified transmissions" and one-way transmissions are not permitted, unless in the following circumstances:

1. In the event of an emergency and demands immediate communication with an station located at a different place

2. For transmitting small test signals or transmitting that are intended for a specific purpose. In these cases the call number of the station should be displayed at least every 15 minutes.

3. In contests or other communications that involve several amateurs

FCC rules

While Amateur radio is as great as it is, it is important to obey the rules that support it. Amateur radio can be a wonderful hobby and a service. But one of the most important obligations of radio operators who operate amateur radio is to follow the rules that regulate it. These rules are outlined within the Federal Communications Commission (FCC) rules and are available in the FCC website.

*The FCC is a government-run agency that supervises amateur radio and other communication systems, such as radio and broadcast television stations, as well as satellite communication systems.

* FCC enforces strict regulations for radio amateurs to ensure that they do not interfere with their broadcasts to commercial radio broadcasting channels.

Additionally, it regulates wireline infrastructures, such as wireless telephone services like cell phones or pagers, Internet protocol (IP) data transmission services, such as DSL as well as dial-up Internet access services; companies that provide cable modem services (like Comcast).

* FCC is also charged with the charge of maintaining computer network equipment makers such as Cisco Systems, Inc. Cellular phones that are used by people rather than organizations or businesses Other types of personal communication devices like two-way radios that are used by firefighters and police officers who are in emergency situations.

Vanity Call System

In 2013 the FCC approved a updated Vanity Call System that allowed owners of General or Amateur Extra Class operator licenses to apply for a specific call number that would be given to them.

Radio amateurs are not permitted in broadcasting broadcasts to the entire public. they are allowed only to connect with licensed operators or conduct experiments in radio communication.

FCC has created the database of all amateur radio call sign codes that have ever been issued and the current amateur radio licenses. The database can be searched using the call number, ZIP code or the first and last name.

Maintenance of mailing addresses

The Commission mandates that transmitting stations outside of the US or its territorial areas to include an international address for transmissions, excluding communications with emergency officials.

Ham operators should keep their address at the Commission not only due to the fact that licensing fees are sent there, but also due to the fact that a large number of unwanted mailings about events at which the operator might be interested in participating often are sent through this method.

Chapter 13: Amateur Radio Operation Procedures

Selecting the Station Frequency

The first step in selecting the right frequency is looking up while listening in the frequency. The frequencies to conduct voice calls are in the range of 1.8 to 30 MHz However, most amateur radio operators are using frequencies that are below 10 MHz.

Two types of communication

There are two kinds of communication, duplex and simplex.

In simplex operation, single-way voice communications are conducted on one frequency. But, duplex system allows two-way voice communication to take place at the same frequency, without interference because each station is equipped with its own receiver and transmitter control.

The frequency range of your radio's signal is limited by its circuitry. Therefore it is best to avoid frequencies that require more power than your device can generate or effectively receive at these higher levels of power (usually about 100W).

It is also important to not operate near other stations which could create interference between their signals , and possibly yours!

A Different Station Calling

Make sure you use the correct call sign when calling a different station. use your first name. It is also important to:

* spell out what you would like to discuss
* Repetition of the call number of the station you're talking to (if they provide it to you).)
* Repetition of the name of your child (if they want it)

For example, here's: "Good morning, this is Joe Smith, W1JSM." If Joe wants to provide additional details, he could explain: "I'm working on a project in my backyard that uses solar energy." When he has explained the project, Joe would end his conversation by saying goodbye, and repeating his call signal repeatedly.

Ham Radio Band Plans

Band plans are made to ensure that hams are able to hear each other and avoid interference. They also make sure that hams can hear emergency services like police and fire departments.

Band plans are essential because they allow you to select the right frequency to ensure that you don't disrupt the other radio amateurs within your region.

Test Transmissions

Test transmissions should be performed at a clear frequency, with a tone, as well as using a digital mode.

In the beginning, you'll need to set your receiver to the appropriate frequency which is 10 meters or MHz. Once you've set it up you can turn on the transmitter and then key it every 2 seconds for every minute (that is 1:02:00).).

The test transmission must be done by using Morse code that is 5 words/minute (5 wpm) with dashes and dots with equal length.

Repeater Offsets

To determine a repeater's offset it is necessary to determine the frequency of the input to the repeater.

You can find out by looking through the Repeater Directory or asking about the local 2-meter or 70cm repeaters.

Once you know what frequency will be used as the input for your local repeater of 2 meters or 70 cm Note it down and keep it in your wallet for use in the future.

How do you select the operating frequency
* Select a frequency that isn't in use.

Beware of using frequencies that are that are not used for other purposes like distress and marine weather as well as aircraft communications. public service events, such as other public-service programs.

* The reason for this is that they could have been financed by groups or clubs like blood drives or public service events such as disaster relief efforts , or public service initiatives organized by groups or clubs like blood drives, amateur radio networks, for instance.

* The only exception is when it is the case that there is no alternative of frequency within the band that is used by that particular type for communication (i.e. voice calls through SSB).

* You should refrain from using frequencies that is primarily a different mode that you might be on the air talking (for example, don't be on 10 meters while the other guys are all speaking CW).

Frequency usage

Radio frequency use for amateur radio is a subject which has witnessed many changes over time. One of the most crucial things to keep in mind is to be a good steward when operating your radio station.

The most popular channels and frequency bands are:

1. 80-meter band (3.5 4 MHz - 3.5 millimeters)

2. 40-meter band (7.0 MHz - 7.3 (7.0 - 7.3 MHz)

3. The 20-meter band (14 - 14.35 MHz)

4. 15-meter band (21 - 21.45 MHz)

Prior to World War I, these frequencies were utilized, however, they are still popular with amateur radio operators due to the fact that they enable long-distance communications using devices that are low in power. They also avoid interference from government or commercial stations operating on these frequencies in different countries or regions all over the world!

CTCSS

Radio amateurs frequently use CTCSS to block their radio transmissions not being played to receivers which aren't supposed to receive them.

CTCSS is "Continuous Tone-Coded Squelch System." It's an approach to encode an audible tone (below the human hearing range hearing) onto an FM radio to enable only radios with compatible CTCSS systems in their scan lists to receive the signal. This keeps nearby repeaters from accidentally keyed up , creating interference. It also can block noise-related signals like those in airports and emergency service frequencies.

The difference between CTCSS and PL

There are two kinds of tones that are used in Ham radio one being continuous Tone Coding Squelch System (CTCSS) which transmits a digital low-frequency signal as well as Private Line (PL).

DTMF

DTMF or Dual-Tone Multiple-Frequency is a digital encoding method for transmitting and receiving data over radio waves, telephone lines as well as the internet. The technology was developed around the year 1970 through Bell Laboratories engineers. The name is derived from the combination of two tones in various ways to transmit information.

Q Signals

They are employed to facilitate communications between stations.

* They serve to mark the end of a broadcast when you are done speaking and wish to return to listening on the radio.

* They are also employed during voice transmissions to indicate how the broadcaster has copied the message and

has a good understanding of it enough not to need to repeat it.

Q signals are typically known as "Q-signal," but some prefer "Q-sig" or "Sig."

Operation in emergency

If you're trapped Do not be afraid to panic. Instead, be at peace and remain as quiet as you can until assistance arrives.

What should you do? The first step is to listen for important announcements from the television or the radio. They could include information on what happened and how to react, and also where you can safely move.

If you require assistance but you are not able to communicate, take note of any further instructions given by emergency response personnel over the radio.

Note: If the earth shakes or another natural disaster you should not be using your radio unless there is a immediate danger of injury or death. ensure that it is tuned to official channels!

Net control and operation

A net in amateur radio network is an informal gathering of more than one (usually at three at a minimum) amateur radio broadcasters using the same frequency for discussion of an issue.

The goal of nets can differ; they could be based on specific regions or interests or may be pure social events that permit members of the ham community to connect with like-minded people.

Operating restrictions during emergencies

The following are some restrictions on amateur radio in an emergency situation:

1. If you're not in the immediate vicinity of an emergency situation, do not utilize the repeater system.

2. Do not transmit via the repeater unless there is some information to share. Incessant key-clicks and ineffective transmissions cause confusion in an emergency.

3. Don't run conversations with your family over a repeater , unless it is necessary to relay information to members of other

organizations and there aren't any other options.

4. Make use of plain language instead of abbreviations, which makes communication more comprehensible for the average person.

Amateur Radio Wave Propagation

Radio waves in all forms are affected by weather conditions. They are affected by multipath, fading or polarization as well as absorption.

Fading

Fading occurs due to the motion of objects in the earth's atmosphere. It's result of reflections, refractions absorption, scattering, as well as dispersion of radio waves as it travels through layer of our atmosphere that is ionized. Fading can also be described as slow-fading rapid fade, and impulse fading:

Slow Fading (also known as Variation or Degradation)

It happens in the event that two signals simultaneously received with different levels of power due to different weather situations (such as temperature variations in the layers above ground). The atmospheric conditions act as lenses, causing delays in the propagation between two signals received from close points on surface of the earth.

Fast Fading (also known as Time Dispersion)

This kind of fading happens when signals arrive from a distant location via several paths that are not in-phase with one another , so that certain parts cancel each and others strengthen each the other.

Fast fading causes rapid fluctuations in signal strength in the duration of milliseconds to seconds, based on the number of paths in the present moment in the transmission

Multipath

Multipath happens when a single signal is received by a receiver through multiple routes. Multipath is caused due to the reflections of radio signals from objects that are in the surrounding. The reflections can cause interference which can cause distortion and fading.

Polarization

Polarization refers to the direction that the electrical field in the radio wave. The direction and strength of the field is determined by the frequency and wavelength. They may vary according to the frequency, weather conditions and many other variables.

The most common kinds of polarization are horizontal (V) as well as horizontal (H). When both kinds are combined, you obtain circular polarization.

Vertical Polarization

In this scenario the electric field runs straight upwards from the antenna's edge up to the base. This kind of signal could be blocked by leaves of trees or tall buildings enough to interfere with the signal.

Horizontal Polarization

In this scenario the electric field is moving from left to right along the path of your radio waves across space when they depart away from the antenna (and/or the transmitter).

This type of signal is higher risk than the vertical polarization to get blocked by trees or buildings due to the fact that it's much easier for trees and buildings to block it in comparison to vertically polarized signals, which move in a single direction and are close to the ground.

Wavelength and Absorption

Wavelength is the distance that exists between two waves' peaks. This determines its frequency as well as its speed of propagation.

It is the process of absorbing energy caused by the wave when it moves through the medium. It is in proportion in wavelength and frequency as well as the medium (material).

Higher frequencies mean that the light can penetrate deeper into materials than at lower frequencies. this is the reason why astronomers are able to look through dust clouds for longer distances with infrared telescopes.

Antenna Orientation

Antenna orientation is a crucial element of the station's configuration and has a significant impact on the strength of signals and even the amount of noise. There are three main kinds of antennas: vertical or horizontal and omnidirectional.

Vertical Antennas

Vertical antennas send and receive signals with the highest quality when they are vertically oriented. This is the reason why AM broadcasters employ towers to help provide support for their antennas that are directional.

VHF and UHF television antennas usually have a horizon view and vice versa. If you're using an antenna with a vertical orientation to transmit, you'll need to position it in the direction of the receiver you want to target and ensure there are no obstacles in between (such as buildings or hills).

A lot of people choose to use verticals to receive because they can pick weaker signals more effectively than other orientations (in plus, being less expensive than high-gain or omnidirectional models).

Horizontal Antenna

Horizontal antenna orientation is well with omnidirectional devices, such as the ham radio transceivers as they were created around sources that are omnidirectional. However should you have line of sight access and your device is pointing exactly at the location you want to location should be perfectly!

Electromagnetic Wave Properties

When an electromagnetic signal travels through an object that is magnetic or electric, the magnetic fields of the waves oscillate perpendicularly to its direction. They are referred to as transverse waves unlike longitudinal waves, which move according to the direction of travel.

Electromagnetic waves do not have mass and move at the speed of light in an atmosphere of vacuum (299,792,458 meters per second).

A wavelength can be determined by the speed at which they travel in space. Longer wavelengths (lower frequency) and slower speeds and smaller distances (higher frequency) translate to faster speeds.

Electromagnetic waves are also distinguished by their polarization. This is the direction that electric fields take to the magnetic field in just one dimensions. They always go towards opposite direction when they are the right angle to one another.

Relationship Between Frequency and Wavelength

The relation between frequency and wavelength is among the most fundamental concepts in the field of radio propagation of waves. The relation between these two parameters could be described in the following manner:

The length of the wave (l) = Speed of Radio Waves Frequency (f)

In other words, the length of a radio signal is the same as its speed multiplied by its frequency.

Speed of light within the free space is about 300,000 kilometers per second, or

3x108 m/s. If we multiply 300,000 by 3x108m/s, we'll get 100 meters/second (m/s).

Therefore, if you have a frequency of one million cycles per second (1 millions of cycles in a second) The wavelength is 100 meters. But if your frequency was just 10 , kHz (10 milliseconds of cycle per second) the wavelength would be 0.01 km! This means that if are a lower-frequency caller as compared to a station that operates with higher frequency, then there's a lower likelihood that they'll be able to be able to hear your call signal even although both stations operate at identical frequencies!

UHF, VHF, and HF

UHF, VHF and HF are the three short variations of the words 'Ultra-High Frequency', 'Very High Frequency' as well as High Frequency. They are the frequency range that radio waves travel at various speeds, based upon their frequency (the distance from one peak and another).

There are many uses for radio waves within the frequency spectrum: television broadcasts and FM radio signals and AM radio broadcasts utilize different frequencies of this spectrum.

Radio waves are typically affected by multipath, fading absorption, polarization, and fading. Fading is the loss in signal strength due to the environment of propagation.

Multipath refers to reflecting radio signals from the objects that are in its path.

Polarization refers to the direction of the radio wave's electrical field in relation to an axis that is defined by the direction of travel (usually the vertical direction).).

Amateur Radio Practices

Amateur Radio Station Setup

The Amateur Radio is considered to be a form of entertainment. It requires the radio transmitter and receiver to talk to fellow "hobbyists."

Amateur radio is also known as Ham radio. The term "ham" signifies the reality that amateurs are usually self-taught, rather than being trained by a school, however it has been employed as a slang term for those who are not trained in anything.

Mobile Radio Installation

When you've purchased the radio, you must to set it up. There are several methods to accomplish this, however the most popular method is connecting your radio with an external source of power and then connect it to a power source to power your vehicle's electrical system. In order to do this, you'll need:

An adapter for a cigarette lighter cable that has an alligator clip on one end and two wires along the other end that connect to the two sides of the electrical battery (if your car has both positive post and negative) or simply an positive one when it doesn't have negative posts.

If you're not sure which direction can be positive for you vehicle look up the owner's manual, or examine the direction that current flows when lights are switched on. If the current is turning clockwise around the post when the lights come on, then the post is positive. Otherwise, it's negative.

* Wire strippers and wire cutters should be part of the basic tool kit you purchase from an electronics retailer but if it isn't equipped with these tools, you can get these tools elsewhere!

An SWR Meter

An SWR meter can be used to determine the power that is reflected in relation to the forward power. This is used to check if your antenna is functioning correctly, and if not working, you will not be able to connect.

How to Utilize An SWR Meter

It's simple!

Simply set it close to the radio to see the number that pops up! If the number is low (say lower than one) it's fine. You're ready for amateur radio practice! If the number is excessive (say greater than one) try tinkering on your antenna until the number becomes nearer to 1 or completely disappears.

How to Connect A Microphone To A Radio Station

Connecting the microphone to radio stations isn't difficult, it's not costly and isn't a big issue.

The most efficient method to connect your microphone to the radio station is connecting it to an XLR cable. To connect your microphone to this type of connection, you'll need:

* A cable that has three wires: two black and two red. The red wires are used to supply positive voltages, whereas the black wire acts for the ground (negative). The cable must be long enough for it to reach the location where your microphone will be located, to where you'll broadcast or record streams of audio.

2 male XLR connectors that plug in channels 1, 2 of your recording desk or mixing device. The connectors are typically color coded in red for channel 1 and white for channel 2 However, this isn't always the case.

How Do I Connect To The Power Source

* Connect your power supply to the radio.

* Make sure you connect your power supply directly to your power source.

* Lastly, you need to connect both of them and connect them to your radio equipment for amateur use for everything to function effectively!

How Do I Connect A Computer To A Radio Station

Connect your PC to radio stations by connecting an antenna cable on your motherboard.

Connect a source of power to your computer for example, the electrical outlet, or a solar Panel battery charger.

Connect your microphone to use it for communications with other stations, by talking into it and then listening for any responses using headphones that are connected to your sound card's headphones connector.

Connect mobile radios (such like walkie talkies) into the ports of the wireless adapter card. Then, place them in the city in order to be able to talk directly with other users via their radios.

Amateur radio allows radio enthusiasts to connect with other people by radio waves. It is possible to use the radio stations of amateurs to communicate with your family and friends, send messages or obtain details from fellow users. Radio amateurs have a variety of communication tools at their disposal such as handheld transceivers (HTs) bases stations, repeaters, as well as satellites.

Basic Amateur Radio Practices

There are a variety of amateur radios on the market today with different levels of difficulty and price. If you are a beginner you must begin by getting familiar with the best way to hook up your radio station prior to moving on to more intricate projects like creating a base station to use for your own personal use or joining a local club for events that are social, such as picnics or field excursions.

Amateur Radio operators can reap immense benefits in taking safety measures when operating these devices as they require electricity to run through them constantly when transmitting signals, which can cause harm if they are not properly maintained

Electrical Principles in An Amateur Radio Station

Before you build an radio radio network, it's essential to brush up on the fundamental electrical concepts, particularly in the event that you're not familiar with circuit analysis, or you're looking to brush up on fundamental physical concepts. The combination of magnetism and electricity can make radio propagation possible. Electrical concepts, and the mathematical research that goes into them are essential to understand the way radio equipment functions. Knowing these principles will allow you to determine the type of measurements that could be performed. These principles are also applicable to other areas that require electricity, for example, electricity generation and electrical engineering all over the world.

Current and Voltage

Voltage and Current are two key concepts in electricity. The term "current" refers to the movement of electrons. Voltage refers to the variation in electrical power between two points.

Current is measured in amps (amps) While the voltage measurement is in Volts.

A Volt can be defined as one joule for every coulomb, or 1 V is 1 J/C. Thus, for instance that you own batteries with 9 volts of charge and two wires that are connected to it, you'll be able to see the that they are flowing with current if you connect two wires. In other words, there's no current flowing as the battery isn't connected with any other thing!

Conclusion

This book is designed to assist you in passing an Amateur Radio Technician test and to help you understand how to effectively communicate using Ham radio. The book also introduced you to some of the fundamental principles that govern emergency communication (EMCOMM). Additionally, you were taught the fundamental concepts of amateur radio , which can prepare you to take the second element of your radio technician test. This test requires a basic knowledge of the regulations and rules as well as electronic theory and radio communication concepts and methods.

Best of luck!

www.ingramcontent.com/pod-product-compliance
Lightning Source LLC
Chambersburg PA
CBHW050400120526
44590CB00015B/1762